ECO-ENVIRONMENT AND GLOBAL BIODIVERSITY

VOL. I

ECO-ENVIRONMENT AND GLOBAL BIODIVERSITY

VOL. I

By

Dr. Reena Mohanka
Department of Botany
Science College, Patna
Patna University, Patna

Dr. M. P. Singh, FMA
Department of Forest Sciences
Birsa Agricultural University
Ranchi-834 006 (Jharkhand)

MD Publications Pvt Ltd
New Delhi

Published by
MD Publications Pvt Ltd
"MD House" 11, Darya Ganj
New Delhi-110002
Tel. : +91-11-41563325, 41562846
E-mail : contact@mdppl.com
Website : www.mdppl.com

ISBN : 978-81-7533-118-1 (Vol. 1)
978-81-7533-120-4 (Set of 2 Volumes)

Price : Rs. 2995.00 (Set of 2 Volumes)

Published and Printed by **MD Publications Pvt. Ltd.** at Times Press, 910, Jatwara Street, Darya Ganj, New Delhi.

FOREWORD

Our future is tied up with the maintenance of the equilibrium of the inexorable forces of nature which brook no interference. The shape of things to come a couple of decade will depend upon how we organise our physical defenses against nature, i.e., in nutshell now we protect our natural resources viz., air, water, land, vegetation, etc.

Eco-environment have profound impact on the socioeconomic life of the people of the country, particularly in areas where people have to depend greatly on forest for their livelihood and also for moderating rapid climatic changes. Apart from the influence of the forests in maintaining ecological balance and preventing soil erosion caused by heavy rain, it affects the climate, particularly rain, wind, heat, cold and moisture. The vegetation and animal (including birds and reptiles) life also get their sustenance from the neighboring eco-environment areas.

This fact has placed the ecology of the planet in a sorry state. Several hundred thousand floral and faunal species were extinguished so far. The present environment conditions are not rendered safe habitat for still more existing species. An environment in which animals and plants become extinct is considered as safe for the human being either.

There is a need to establish a new kind of ethics viz. "Development without destruction and production with protection" then it would be possible to synchronize ecological activities with the world of commerce. It is a high time to re-introduce symphony in the biosphere. While planting trees one should must consider about the endangered species of flora and fauna. Trees should be grown not as a single species but in an association with grasses, creepers and others co-existing living beings and consideration of natural local habitat species. This will form a symphonic biological matrix in a given habitat. However, it needs a great ecological wisdom, long site and well being of animate.

The authors have accomplished a difficult task in producing a fairly detailed and up-to-date book which will be very useful to the scientific communities as well as concern people.

The publication of a new Eco-environment & Global Bio-diversity textbook is a rare event. This one is worth the wait.

Saim Parioda
Chairman
National Knowledge Commission
Govt. of India

PREFACE

Our father always felt uncomfortable about the quality of the river water, although his faith in the divinity of the Ganga was limitless. He was of course, ahead of his times, for the deterioration of environment was hardly a social issue in the 1960s. Today, the nature & cause of problem is well recognized, and its gravity is well understood. While the government searches for solutions, the people are learning to distinguish between the divine powers and the physical powers of their natural resources.

Mother Nature is an Integral part of the Indian culture and heritage, In Hindu mythological stories, rivers and Mountains emerge as angels, and air and trees appear as human beings, being able to feel, hear, and talk. Despite this "humanization" of the environment, the abuse and destruction of natural resources has occurred at an alarming rate.

It is necessary to introduce symphony in the biosphere and establish a new kind of ethics, viz. "Development without destruction and Production without protection". The quality of life can be enhanced by improving the quality of air, water, land and land resources.

The global environmental issues of deforestation, desertification, land degradation, deterioration of water sheds and floods, pollution, erosion of biodiversity have to be tackled on priority basis. Efforts are being taken on Global basis such as "Convention on Global warming", "Biodiversity Conservation", "Non legally Binding Authoritative statement of principles for management, conservation and Development of forests", "Earth's Summit's Agenda 21" and several other declarations.

In the Indian Constitution, several provisions have been given to protect environment and safeguard the natural resources by adopting a New Forest Policy (1988), Water Policy (1987), Housing Policy (1988) and Land-use Policy (1987). It has also adopted "The National Conservatory, Strategy and Policy Statement on Environment and Development (1992), Policy statement on the Abatement of Pollution (1992), Environmental Action Programme (1993) and Plethora of Legislations".

Till now though studies on the forest and forest resources have been done by some thinkers, all aspects of the same have not been covered fully. The present book alms at covering all aspects of forest in relation to their contribution to the economic welfare of the people of India.

The book attempts to integrate our concerns for conservation and development of Eco-environmental protection, poverty alleviation and human welfare. We hope that this book will help in creating Eco-environmental awareness and in the implementation of numerous Eco regeneration projects.

A number of experts have contributed to the development of this book. Their contribution is gratefully acknowledged.

Reena Mohanka
M. P. Singh

CONTENTS

1. Introduction 1
2. Natural Resources 12
3. Community Ecology and Pollution 46
4. The World Environment Scenario 84
5. Pollution Control 113
6. Environmental Impact Assessment 144
7. Air Pollution 195
8. Water Pollution 235
9. Noise Pollution 259
10. Domestic Pollution 273
11. Atmosphere 285
12. The Environment (Protection) Bill, 1986: A Bill 313
13. An Introduction: Biodiversity 331
14. Biodiversity Loss 340
15. Technique of Biodiversity Conservation & its Implication 358
16. Biodiversity Conservation: India Scenario 412

1

INTRODUCTION

The impact of human activities on the physical and biological environment has a consequential effect on man and his society due to the changes in the environment, illustrations of such changes occur in the archaeological records, such as the gradual desertification of parts of Sind and Rajasthan which one sustained larger populations and greater agricultural activity as well as more extensive grasslands. A dramatic evidence of such reciprocal interaction in our times is the discovery of mercury poisoning cases in some areas of Japan through eating of fish contaminated by mercury. For a few years the effect, or disease, in man was not even diagnosed as mercury poisoning but as a new illness. Later, it was traced to the mercury present in the fishes. Normally, mercury is only slightly toxic. One can ingest a few grammes of mercury without severe effects. However, research revealed that the high toxicity was due to poisonous effect of mercury called methyl mercury (CH), (Hg_2). These were being produced by certain anaerobic micro-organisms in the sea bottom of shallow areas of the east-coast of Japan and were being ingested by fishes along with sea weeds which constitute their normal diet. The physiology of fish was such that the poisonous mercury compounds tended to accumulate in some of the organs of the fish, the source of mercury was a product of industrial activity of pulp and paper mills which discharged small amounts of mercury in their effluents. The entire marine and human food chain is complex in which the microorganisms of the sea bottom play an essential role to convert mercury and to link it with the food chains since the microorganisms enter the marine food chain and work upwards to the edible marine fishes and ultimately to the human population who eat these fishes. Since methylated mercury are not easily eliminated by the human body, they tend to accumulate in the body, and over a period of time result in human illness and death. The diagnosis of the cause of the so-called Minamata disease and consequent preventive steps took some time as investigations over a period of years revealed the true nature and cause of disease and source of the pollutants.

Nature's balance provides other microorganisms which break down the methyl and dimethyl mercury to mercury so that the concentration of the methyl and dimethyl mercury remains low. Recently, the presence of mercury has been found in the fishes in the gulf of Cambay and in the coastal fishes in certain parts of the Bombay coast. The increase of the mercury is still unidentified.

Less dramatic but, equally if not more disastrous is the erosion of topsoil in the submountain slopes by floods and landslides preceded by and due to loss of vegetative cover resulting from tree felling (for fuel and logging) and loss in the process is estimated to be 105 mha in India. This is due to paper mills effluents in many areas of the country, the acid sulphite process to separate the lignin from celluose leads to formation of the lignosulphonates, these lignosulphonates are less biodegradable that the lignins are more toxic. Their presence in water affects not only fish and aquatic fauna but also downstream users of water because of smell, toxicity, colour and decreased utility human and agricultural use. Pesticides* like DDT have been found in human beings in autopsies in Delhi and Bombay and in animals at several places because of their entry into the food chain in a variety of ways.

There are other well-known cases like the release of a toxic chemical (dioxin) from a chemical factory by an explosion in the factory in Seveso near Milan in Italy and its entry into the surrounding country and subsequently ingested by men and animals in the neighbourhood. More recently in Michigan a (polybrominated biphenyl) in bags normally used for packing entry of these compounds into the food chain through cattle feed affecting both cattle and human beings. It was first neurological disorders and skin diseases in man which was traced due to the presence of the chemical. In Delhi and Punjab DOT has been found in cattle and human beings, children and young people tending to show greater amounts in their organs than older people.

There are many important activities of man apart from such accidents, industrial effluents and wastes, which have an impact on man or not the ecological system as a whole. The large scale of coal and petroleum as fuel for energy produces vast amounts of carbondioxide (CO_2). The amount of CO_2 has been found to increase from 300 ppm to 400 ppm. It is thought that such increases, if continued, could result eventually in global climatic and temperature changes, there are instances of rapid deforestation of hill slopes leading not only to reduced release of oxygen from plants by photosynthesis but also increased landslides and floods in the plains down below in U.P., Bihar and Assam. There are increasing deserts in Rajasthan and Sind due to overgrazing of marginal lands at the edge of deserts, leading to the removal of green grassy cover of the land and decrease in the moisture retention capacity of the soil. This has resulted in decrease in rainfall. Consequently, the land once marginally habitable has become desertified.

Technological activities of man have increased energy consumption. Coal and petroleum are the main sources of energy today. In mining coal and extracting other minerals from under the earth there is often subsidence of land and holes and gases on the surface of the earth, leading to the accumulation of water in these areas, as well as changes of water courses and runoff affecting the ecology of these areas. It also gives rise to main environmental problems by local flooding, abandoned, water logged lands and increase in insect and waterborne diseases. The impact on the health of the community or of ugliness of the countryside is not always quantifiable.

The excessive use of fertilizers and pesticides contributes substantially to the ecological disturbances and environmental pollution. Excessive use of nitrogen fertilizers, for example, can lead to the leaching of these excess fertilizers into the water bodies.

These can be transformed by microorganisms into the nitrites and carcinogen as nitrosoalines and can find their way back to men or animals through the food chain.

Excessive fertilizers can also adversely affect the nitrogen fixing organisms or microorganisms in the soil and reduce the natural nitrogen absorption and production of soils, similarly pesticides, besides killing the unwanted pests, insects or birds and can enter the human system through food are contact and become a health hazard. They can also indirectly come into the human system through animals, which ingest the plants. Other human activities also cause various ecological problems; for instance, over-grazing in a grassland can denude it to the extent that it is gradually transformed into a desert. In Rajasthan, it is estimated that the desert in certain areas is advancing at the rate of half a kilometre a year.

Many of the large-scale human activities-industrial, agricultural or transport tend to alter the natural balance of living organisms in an ecological system. Such a disturbance in the ecology to the system can affect human societies in a number of ways both in the immediate and in the long-term. A grassland can be denuded and gradually transformed into a desert. Denudation of forest and grassland vegetation can cause long-term climate changes by reducing rainfall and thus perpetuate and strengthen the desertification process over decades and sometimes over centuries or millennia. The courses of streams and rivers can also destroy the ability of trees, alongwith the soil and humus which support under shrubs, microorganisms, insects and animals which live in various symbiotic relationships with each other and the larger plants and trees. Once the forest is gone, these animals, plants and other organisms are also destroyed, making regeneration difficult or impossible. Larger trees and animals tend to be more valuable and are often the first to disappear.

Man is largely, not wholly, dependent on his environment for the resources that sustain and give him various products. In extracting these resources whether by way of agriculture or minerals or obtaining energy resources, the topsoil may be over-exploited leading to dust bowl-like conditions or so scared as to be unfit for use. Forests may disappear if trees fall faster than replacement. This leads not only to denudation of topsoils but there may well be no fuel for the poor and no paper for the schools. One of the primary concerns of environmental management is the preservation of natural resources base and wherever possible enhancing further its productivity in the long term. Unfortunately, the rapidly increasing population which has been already doubled since independence. It is likely to exceed 100 crores by the end of this century. This large population needs food, energy, shelter and other amenities of life and in increasing amounts. The need of land is roughly proportional to the population if other things remain the same. However, the 150 million hectares of current agricultural land is not likely to increase very much. On the contrary, urbanization, developmental projects, roads etc., are going to encroach more and more on agricultural land and such land will be increasingly precious.

In various ways, therefore, the ecological system and the human environment are closely connected. These linkages tend to confuse us sometime, because the words are often used synonymously. Environment in the context will be used specifically from the human point of view. In some ways this leads to a larger concept which also takes into account human interactions with an ecological system and the mutual effects of human activities on nature and nature on man. When primitive man lived as a part of the ecological system, he was one of the many species of animals which interacted with one another and with the physical world. Civilized man has gradually, through the use of technology, transcended these symbiotic relationships to a considerable extent. This has given man a singular ability to interfere with or even damage nature, not always with wisdom or with advantage to his own long-term future.

Profound alterations of our environment are now being caused by the growth of modern technology, particularly during the last 100 years and at increasingly accelerating rates at present. Such alterations of our environment have been speeded up by the singular growth of human population at an almost exponential rate over the last 250 or 300 years, doubling almost every twenty-five years or thirty years, part from the population growth, the other singular factor of times is the many-pronged development of technology and the expectations that this has created in the process of techno-economic development of all societies including India. The expectations of the people call for progressively greater availability of goods and send us through process of development. However, development requires utilization of natural resources. If the use of these resource degrades the remaining resources for future use and depletes them in a manner that these are lost to future generations, and governments can be accused of being shortsighted, selfish and greedy. The activities of man are thus compounded by factors of increasing technological ability, energy and material consumption and large population increases. The environmental impact of such increasingly global in character. It should be concerned not only about the disturbances to the ecological systems, but increasingly concerned seriously, in the long term, about the future of planet. Man has now the capacity to destroy the entire living systems of the world for the first time in human history.

The close interdependence between resources, development, population and environment seem obvious on examination. Unfortunately, quite often in the process of development, planning and the utilization of resources it tends to ignore the factors of environmental risks and population pressures that are evolved in the development process. There are other social factors, neither articulated clearly nor taken into account, such as discounted resource base estimation which give a framework for calculating what is happening to our resources base with time.

As human population continues to increasing in the next few decades, the need of adjusting themselves to nature and of not disturbing ecological systems significantly will become even more necessary. The impact of human activities on the environment has to be taken into account and where necessary, ameliorated by suitable measures, so that human activities do not become self defeating in their not being able to create human well-being which was intended. Under increasing material and energy use man's relationships with his environment will become exceedingly important and complex. Such use of materials and energy will usually lead to stresses in the environment.

The policy will have to be carried out in term of actions and resources allocated for activities. Its execution starts with the education of children and adults and includes monitoring, investigation, research, enactment and enforcement of social and scientific groups which concern themselves with environmental question. The total package of policy, low education and activities is the task of environmental organizations. Environmental management can exist at several levels. Briefly they can be institutional, industrial, civic, cooperative and governmental at local state and federal levels, depending on the nature and site of the task of issues.

The growth of modern technology and industry has increasingly helped to push up the population in urban settlements. In urban areas, the dynamics of population increase under the impact of modern technology. In the last two centuries have been to encourage large numbers of people to live close to industries and factories and to each other. In the last hundred years or less this trends has been served equally effectively by people tying themselves to transport systems. Industries and factories have tended to be created at convenient points where power, raw materials, transport and communications were easily available. This has led many industries to locate themselves close to each other, both for access to raw materials and transport and for the mutually supporting roles that industries tend to have with each other. The dynamics of industrialization and population increase have led to the growth of the large modern urban complexes.

The two situations-urban and rural are thus completely different. The urban settlement is sometimes called urban ecosystem, in which natural linkages have been significantly or largely destroyed or perverted and a new system of social and economic linkages have developed. The urban complex as a consequence has developed new sociological, psychological, and technological parameters. The urban settlements are, thus, not only artificial in the sense that they are man-made but they have also different set of interactions and dynamics and do not fall strictly within the definition of ecological systems. Urban ecosystems is however, an increasingly favourite terminology to describe the urban systems. The rural complexes on the other hand, still remain closely related to their environment and the linkages with the natural ecosystem are not entirely disrupted. While rural settlements also have problems both psychological and technological in nature these have yet to take acute forms. Part of the reason is perhaps the safety value in rural complexes tend to be the resources base, its optimal utilization, the various options on utilizations, extension of the resource base, the impact of increasing population which the resources base has to sustain and the surplus create for urban centres.

There are social aspects of rural areas. With the increasing local perception of deficiency of agricultural, educational, employment and cultural opportunities, rural dissatisfaction has increased and urban migration has accelerated. Such factors underline the perception that the quality of rural life is unsatisfactory. Trends of urban migration have accelerated particularly as rural population increases have brought more pressure on a fixed amount of land tending to reduce per capita income. In such migration out of rural areas, there are both rural and urban problems.

Thus, a whole range of potential and actual environmental problems ranging from small perturbations in undisturbed ecological systems into which man is now making ingress, such as introduction of new plants or animals into natural forests and natural lakes to entirely man-made human settlements such as cities and industrial complexes which are complex and artificial environmental systems. In man-made human settlements, man is in a sense primarily responsible for looking after the system to ensure and improve present human well being. Man is also in sense responsible to see that the resources base is not unduly jeopardized or badly exploited to the detriment of future people (our children and grandchildren) who will live and work in these human settlements.

Many general principles of management of any system apply equally to the management of environmental systems. However, the specific scientific, technological, social and biological aspects which relate to the environment, have also to be incorporated in the management of such systems. Some of these factors such as the effects of major engineering or technical activity may not be fully known or understood. They pose problems to which no ready answer are possible at the time that action is contemplated. There are aspects of uncertainty and risks which cannot be avoided when decisions have to be taken and action contemplated. There are many kinds of environmental systems and each has its own peculiarities and managerial problems. For most management purposes, a broad classification of the various environmental system is often adapted: (a) biosphere reserves, natural lakes or natural forests which are desired to be reserved for purposes of gene pool preservation, species protection, scientific studies such as these of evolutionary processes, aesthetic of entertainment purposes and the protection of geological or natural hydrological structures, (b) rivers and river valley forest used as resources of timber and fuel, hill sides, beauty spots etc., (c) agricultural land, agricultural waters, dams and artificial lakes, and (d) urban and industrial complexes and other highly disturbed zones such as mining areas.

Similarly, dealing with the problems of a moderately disturbed ecosystem such as a river, the problems of fresh water fauna, the utilization of the river water by people who live near the banks further downstream, the problem of human settlements on the bank of the river at the location and downstream, and the steps to prevent floods have to be looked at very carefully both in terms of aesthetic value of the natural river banks and the economic utilization of the river. The conflict inherent between maintaining the natural beauty of the two banks and the building of roads and parks or human habitation (including expensive hotels) on the banks will make competitive claims into which the environmental management system has to be looked closely. It has to approximate to achieved the clearly defined initial objectives with which it was entrusted. Social and economic pressures that developed in the meantime will, in addition have to be taken into account to prevent serious distortion.

As far as highly disturbed or rather the man-made environmental systems such as cities and farm, dams and roads are concerned, the problems are more acute. There is usually a greater clarity regarding water is desired even though this clarity may not apply equally to state how this is to be done. If a dam is to be built or a water reservoir has to be preserved one knows that within the precipitation basin of a dam or a reservoir land erosion should be minimized so that the reservoir does not get filled up with silt in a short time, the problem often is how felling of trees or forests or other ground cover is to be regulated within the existing social parameters of energy needs and social habits of the people who previously depended on firewood or on the grazing needs of sheep, goats or cattle by nearby communities within the overall objectives of the projects.

Methods in applied ecology using simplified models continue to be developed to provide some theoretical base to understand the complexity of the system and suggest answers to policy questions such as possible fishing quotas or mesh size of fishing nets or spraying to control agricultural pests and insect born diseases such as malaria or use of biological agents for control of pests. The sample models do not provide a detailed and complete theoretical structure of ecosystems or even an understanding of the complexity of internal relationships. The availability of large computers have made it possible to incorporate increasingly complex and interlinked elements in a large system of equations to simulate more complex system for purposes of study. These cannot claim in any sense to be complete studies but can in certain regimes or circumstances elucidate some understanding of the more complex behaviour of ecological systems.

2

NATURAL RESOURCES

Today we get 7 times more electrical power from 1 ton of coal than we did in 1900. Similarly the energy needed to produce 1 ton of pig iron has fallen eight fold since 1800. In the United States in 1900 only 10% of the copper was recycled. Today about 40% is recycled. However, while many matter resources such as copper, lead and silver, can be recycled, we can never recycle energy resources.

Once a fossil fuel resource, such as coal, oil or natural gas is burned, it is gone forever as a useful energy source. The concentrated energy in the fossil fuel is released as heat, which is eventually dispersed into the earth's atmosphere. From there it flows back into space. Sometimes technology can solve the problem of a scarce resource by finding a substitute or replacement. For instance as a structural material bronze replaced stone; iron replaced bronze, steel replaced iron and now aluminum and reinforced plastics are replacing steel for some structural uses. As an energy source, animals replaced human muscle power, coal (to produce steam) replaced animals and oil replaced coal for many uses.

There is a vigorous debate over whether a combination of coal derived energy and nuclear power (obtained from Uranium) or a combination of solar, wind and plant (biomass) energy may soon replace petroleum and natural resource.

In addition to technology resource use is tied to economics. Something is useful as a resource only if it can be made available at a reasonable cost. For example, once we deplete the easily available supplies of a resource, we have to look harder and dig deeper to find remaining supplies. In the costs of findings and making a scarce resource available rise, the resource will eventually become too expensive for most people. Higher costs may stimulate a new search for new supplies or make mining and processing lower grade deposits economically feasible. But regardless of what we are willing and able to pay, we can't get a resource out of earth if it is not there.

There can also be an economic limit to recycling. Typically recycling is cheaper than mining virgin materials but only if the material to be recycled is not too widely dispersed. For example, if products made from iron and steel, such as cars and toys are thrown away, buried and widely scattered through use labour and energy costs for finding and collecting the objects may be too expensive to make recycling feasible.

The continuity of a resource may also depend on the impact, its mining, processing, and use, has on the environment. Even if affordable supplies of a resource is available, its use (at least for certain purposes) may have to be abandoned if this use seriously threatens human and other forms of life. Sometimes the environmental effects of a resource use can be minimized and cleaned up. But this cost money and in some cases the cost may be so high that we can no longer afford to use the resource.

The sum of all physical, chemical, biological and social factors which compose the surroundings of man is referred to as environment and each element of these surroundings constitutes a resource on which man draws in order to develop a better life. Thus, any part of our natural environment such as land, water, air minerals, forest, rangeland, wildlife, fish or even human population - that man can utilize to promote his welfare may be regarded as a natural resource. Ranade defined a resource as a form of energy and/or matter which is essential for the functioning of organisms, populations, and ecosystems. In the particular case of humans, a resource in his words is any form of energy or matter essential for the fulfilment of physiological, socio-economic, and cultural needs, both at the individual level and that of the community.

The functioning of ecosystem or of technological civilization involves a continuous flow and consumption of energy, sun being the source of it, as regards ecosystem and fuels, as regards technological civilization. In the light of this fact, a resource may be defined as one of the various forms of primary energy present in the nature.

The five basic ecological variables-energy, matter, space, time and diversity are sometimes combinedly called natural resources. As such, laws which govern changes in these quantities have great applicability in the resource use.

The Classic subdivisions of resources include renewable and non-renewable resources. Renewable resources are those which are reproducible and are obtained from the biomass of living organism, e.g. forests and fish. Resources which are not reproducible and are obtained from the finite non-living reserves are called non-renewable resources e.g. coal and metals. Solar energy although having a finite life, as a special case, is considered as a renewable resource in as much as solar 'stocks' are inexhaustible on the human scale.

Some authors prefer to classify resources, into biotic (or living) and abiotic (or non-living) resources, forest, agriculture, fish and wildlife falling in the first category and land, water minerals etc. in the second.

Quantity, mutability and reusability are the attributes on which this classification is based. Inexhaustible and exhaustible are main categories of resources based on their stocks in nature. Inexhaustible resources are further divided into two groups. Immutable and mutable resources, depending upon the possibility of their qualitative degradation as a result of man's activity. On the other hand, exhaustible resources, being limited in occurrence, are vulnerable to both quantitative and qualitative degradation. Thus their availability depends upon the method of use, so these are classed into maintainable and non-maintainable resources. Supply of maintainable resources could be made to last for long through wise use, however, some of them have capacity to reproduce called renewable, while other cannot reproduce called non-renewable resources. Non-maintainable resources have static supply and when destroyed or consumptively used are irreplaceable, yet some of them have potentiality for reuse called reusable resources, and rest whose permanent exhaustion certain are called non-reusable resources.

The term "conservation" to many means "hoarding" or to control supply of goods in such a way that some part is left for the nature. Most people interpret conservation as total protection or restriction in consumptive use of resources, therefore, wildlife conservation to them means a total ban on killing of any animal and the forest conservation means a complete moratorium on all feelings the term, in fact, is derived from two Latin words-con (together) and server (to keep or guard), thus meaning 'to keep together'. The dictionary meaning of the word is an act of preservation. The term was proposed by Giffor Pinchot in 1988, who conceived it from the title of 'conservares' given to British officials in India appointed to look after the protection of natural resources.

Prof. H.M. Rose defined conservation as the optimum allocation of natural, human and cultural resources in the scheme of national development, where by maximum economic and social security will be assured. In 1967 the American President John F. Kennedy looked upon conservation as "the wise use of our natural environment... The prevention of waste and despoilment while preserving, improving and renewing the quality and usefulness of all our resource". World Conservation Strategy declines conservation as "the management of human use of the biosphere so that it may yield the greatest sustainable benefits to present generation, while maintaining its potential to the needs and aspirations of future generations". Thus the conservation is a practice embracing preservation, maintenance, sustainable utilization, restoration and enhancement of the natural environment. Summarily, thus the aim of conservation is two fold: (i) preservation of quality of environment, and (ii) to ensure continuous yield of useful material living or non-living by establishing a balanced cycle of hardest and renewal.

The concept of resource conservation has now come under a still wider field on environmental management which is, in essence, the process of allocating natural and man-made resources so as to make optimum use of the environment in satisfying basic human needs at the minimum and more, as far as possible, for an indefinite future, and at the same time preventing depletion and degradation of resources.

Environmental management being a too vague, general, or trivial subject has not developed as a full-fledged discipline so far, further, in many developing countries it is viewed as being irrelevant and counter productive, while in advanced industrialized countries it is often interpreted as a means of coping with the 'burden' of environmental legislation and problems put forth by 'overzealous' environmentalists what to talk of under developed world! The global population, however cannot and should not overlook and deny the last deterioration of environmental world over resulting in dwindling natural resources, pollution of air, water and land, extermination of a number of life forms, down going healths status, danger to cultural assets and many socio-economic backlashes. The root of the problem lies in too many people using too many resources wastefully.

Since the 'total natural resources available are limited, the way out to the problem seems only through control of population and also the per capita consumption of resources. Because it is difficult to bring down per capita consumption, the recycling and stringent conservation can bring hopes for sustained development without reducing per capita use of resources.

Land

Soil consists of the weathered layer of the earth's crust with living organisms and products of their decay intermingled. Lying on the rocky core of the earth", this complex mixture is one of the four prime requisites for life. Along with sunlight, air and water, soil supports all living beings on land, plants as well as animal. Altogether terrestrial plants have their roots in soil from which they absorb life sustaining water and nutrients. Man and all land animals, in turn, depend on these plants for food directly or indirectly. Foods, vegetable oils and fats, forage for livestock, fiber, wood and various other indispensable agricultural products come from soil. Thus, a nation's economic well being is inextricably linked with the fertility and abundance of her soil resource because of mounting demands of swelling population and long mismanagement would put in jeopardy the very survival of man.

Land-use and abuse

The land although appears to be an unlimited resource but, in fact, its injudicious use would limit the availability of this indispensable life-support system. Thus, a land-use may be poor or good, if in a region of low rainfall grasslands are replaced by wheat crops, a temporary desert will eventually result; or if the human habitation and factories are developed on flood plains, the damage is almost inevitable. These are the instances of poor land use. Conversely, if the grasslands are used for moderate grazing by livestock, and if the flood plains are used for forestry or agriculture, these will imply good land use. Ecological principles have great applicability in land-use planning. Classification of use should, therefore, be based on natural ecological features of the land in question viz, soil type, slope, biotic communities etc.

Rural land-use planning

To a farmer, soil is the essence of survival and his economic well being is firmly linked with the fertility of his land. In rural land-use planning, concentration is chiefly devoted to creating and developing more farmland by removing forests and natural grasslands, drawing marshes and swamps, channeling streams for irrigation, and so on. Unfortunately, no effort is made to save existing prime farmland from degradation by ill-planned development. Rather than exclusive farm, range or forestry oriented planning, the land-use should be made under broad consideration of social, economic, and ecological intricacies.

Urban land-use planning

Deterioration quality of urban and suburban environment is to a great extent the result of injudicious land use and is a threat to the whole socio-economic system. Thus, planned cities are as necessary as planned farms. Although, the implementation of land-use plan is infinitely more difficult in urban areas than in rural ones because of the human problems involved, yet if not heeded it would lead to chronic pollution, epidemic, and social disorder. In any urban land-use planning it should be seen that: (i) at least a third of all land remains as protective open space in the form of parks, green belts, or refuges etc., including adequate space for efficient treatment of domestic and industrial wastes, (ii) residential housing around villages or cities should be developed in clusters with each unit separated by broad green belts, (iii) typical topographic situation like stream, valleys, steep slopes and marshes etc., should be retained free from houses and other high density uses; and (iv) waste disposal areas and industries should be kept apart from human habitations.

Management of land

From the above account, it is clear that major problems emerging out of abuse of land in cities include unsafe and unhygienic dwelling and those in rural areas are of erosion of soil and loss of fertility. While urban problems can be prevented and cured by proper town planning and municipal facilities, the problems relating to rural areas need long term soil conservation programme.

Prevention and control of soil erosion: The primary purpose of soil conservation is to prevent soil erosion and heal the damage where it has not advanced too far to respond to curative methods. This is possible largely through two means:

1. The land should wear a vegetational cover throughout the year. Plant roots, bind the top soil and prevent it from being drifted so easily with water or air. The vegetative cover also slows up runoff water and causes much of it to percolate into ground that promotes more vegetative growth. The vegetation whether grasses, trees, or cultivated crops should fit to the capabilities of the soil and water available.
2. Engineering and agronomic practices should be applied conjointly to counteract the erosive force of both water and wind on the soil. These include contour cultivation, cover crops and crop rotation, and contour strip cropping, used alone or in combination with terracing where it is required. When terraces are properly constructed and adequately supported by approved cropping and tillage practices, they prove to be one of the most effective, erosion-control measures applicable to cultivated sloping lands.

Reclamation of eroded lands: Conservationists are divided into, two camps over the issue of reclamation of land already arrested by ravines. One camp of conservationists and naturalists believes in the policy of "forgive and forget" and advocates that ravines should be allowed to exist, while the other feels that efforts should be made to regain the lost land through buldozing or some other means. If the ravines are left in there present state, these will spread further and engulf more of productive land. While the violent means to regain the ravines in their near original condition may provoke numerous agencies of nature to take revenge on man. Thus, following considerations are important in negotiating the erosive land:

(i) Ravines should be provided with sufficient and suitable vegetative cover.

(ii) Instead of agriculture, these lands should be reclaimed for forestry, pasture, or horticulture.

(iii) Their deficiency in nutrients and moisture for plant growth should be improved.

(iv) Further misuse of such land should be prevented. Over-trampling by man and cattle should be avoided.

(v) Past damages should be repaired and recompensed under strict ecological consideration. For instance depressions may be filled up, slopes eased out, cattle, trials fenced etc.

(vi) Vegetative cover provided should be protected against reckless destruction by local population.

Water

The origin of water on the earth is not clear so far. However, the current presumption is that the primordial earth had no oceans, and perhaps very little atmosphere. It is believed that the volatile constituents bound in the earth's crust, oozing the surface through volcanoes, rock movements, and hot springs, condensed to form the ocean and the atmosphere. This way, perhaps this remarkable combination to hydrogen and oxygen-called water came into being and eventually became an indispensible component of the earth's environment.

Water moves from the ocean to air to land to ocean in set cyclic pattern called hydrologic cycle. In other words, this cycle represents a continuous process of exchange of water from the earth's surface to the atmosphere and vice versa under the influence of solar heat.

It is estimated that of the total available water 97.3 per cent is contained in the oceans and out of remaining 2.7 per cent most is in solid form. The amount of water actually available in the atmosphere is a very minute fraction and estimated to be only 1×10^{-5} of the world's total water.

The factors which control the process of evaporation and evapotranspiration include temperature, humidity, and wind. It is a continuous process going on day and night from the ocean surface, ground surface, inland waters (like lakes and streams), plant and animal surfaces. Rising in the atmosphere this water forms clouds that float around in gaseous or droplet form. The atmosphere above the earth up to varying heights of 10 to 17 km (depending upon latitude and season) is called the troposphere up to which the clouds float and in which all weather changes take place. Water returns to the earth in the form of precipitation, most of which takes place over the ocean and a little on the land surface. Rainfall on the land surface may have following fates: it may percolate through the soil and become a part of ground water regime; accumulate in a pond, lake or reservoir as inland surface water; flow down to the ocean in the form of stream; taken up by plants and animals; or evaporate.

Water Requirements and uses

Goethe rightly said that, "Everything originated in the water, and everything is sustained by water". Water is needed to fulfil diverse requirements in so many diverse ways. It is vital to life since for all physiological activities of plants and is essential for animals as well. On an average it constitutes 80 per cent of protoplasm ranging from 8 per cent in dry dormant seeds to 95 per cent jelly fish. It serves not only to quench the thirst but also to meet the food requirements because it is an essential raw material in the process of photosynthesis through which green plants make food that is used by all trophic levels directly or indirectly. Ambient humidity which is indeed the invisible form of water that surrounds us is necessary to prevent desiccation of terrestrial life forms. Thus, it can be said that water is a biologically essential and it serves as milieu internal as well as milieu external.

Flood plains of rivers have been the cradle-of civilization and centres of population since ancient times. The reason being their high fertility an easy accessibility to water for various uses like drinking, washing and transportation. Egypt was developed on the flood plains of the Nile, Babylon on the flood plain of the Tigris, and North Indian on the flood plains of the Ganges. Even today the Mississippi plain is about the most densely populated area in the USA as flood plains of the Rhine in Europe, and the Yellow river in China.

Besides temperature, water is the other such key factor that influences the global ecology. Covering about 70 per cent of land surface and circulating in hydrologic cycle water influences weather and climate on any region and thus its flora and fauna.

Agriculture is dependent on water, may it be rainfall, or surface or ground water irrigation. It eventually is a more important commodity in a country like India because her largely agrarian economy depends critically on water, huge amount of protein rich food in the form of fish, Shell fish, and prawn is obtained from the inland as well as it is very potent industry.

The kinetic energy of gravitational flow of water is tapped and transformed into electricity, what is called hydro power. This concept of electricity generation is one of the oldest and cheapest tricks largely used world over. Moreover, it leaves no pollution-burden on nature as in case of thermal or nuclear power generation.

Hardly any industry can do without water. It has multifarious application in industrial processes viz., a raw material, solvent, chemical reactant, coolant, and cleaning agent.

"Water, water every where not any drop to drink is the situation not only found amidst sea but also at places where industrial and domestic refuse make the benevolent water sources so severely contaminated with pathogens and or toxins that they are lift impotable. Being the best dilution medium it is frequently used as the cheapest and convenient way for liquid and solid waste disposal. Further it was practically impossible to maintain sanitary conditions without water as it is needed in all common cleaning and washing practices.

Water at many places serves as a very convenient means of navigation, besides, it adds to the aesthetic value of the landscape and provides opportunities for recreation and sport.

Water being unevenly distributed over the land, influences interstate and international relations aid thus is the cause of what is known as hydro-politics.

It can, therefore, be said that while water is essential for life it is significantly linked with social, economic, political and ecological intricacies.

Problems and management of water

Management of water implies making the best use of available water resources for human benefit while not only preventing and controlling its depletion and degradation but also developing it in view of the present and future needs. Water, like forest, is a multipurpose resource and it is important to see that its various uses should not conflict with each other and it can be enjoyed in its totality by man and others. Thus, its right allocation, and quantitative and qualitative conservation are the primary tasks before water managers.

Floods, droughts, improper use, pollution and disease transmission are the important problems related to water. Because usable inland water resources constitute a very minute fraction of the world's total water and because man largely depends upon this minute fraction of water for his various activities, it is important to make efficient use of available water and at the same time prevent abuse of it. This has to be considered not only in giant irrigation projects or industrial and municipal processes but also in every day life, taps running unpurposefully, leaking water supply pipes, using filtered water for gardens and building construction are a few instances of inefficient and misuse of water in daily life.

Water being the best and convenient dilution medium is the worst affected resource by pollution. To a common man pollution of water means only the drainage of industrial effluents or domestic sewage in natural waters. It may astonish a lay reader that even pure distilled water when released in sea may act as a pollutant since it alters the normal salinity of sea water which adversely affects the marine life. Besides chemical alterations, changes in physical condition of water also have grave effects on the biota of system as during thermal pollution caused by the discharge of heated effluents in a body of a water. Pathogens, chiefly entering natural waters through municipal wastes rich in excreta and other organic filth, also render the water unsuitable for human use.

In fact, it is not only, the man who has to use the water resources by these are equally important to land animals and more to those fish and other aquatic animals who live in it. Thus pollution of a body of water does prove detrimental not only to the human society but also to regional ecology and the ecology of very aquatic ecosystem. It is, therefore, essential that the water should not be treated as a simple repository of waste disposal, and if inevitable, the wastes only after proper treatment should be released in it under strict ecological consideration.

The qualitative degradation of water is also in a way quantitative depletion of usable resources. Therefore, recycling of waste water after due treatment would relieve the water scarcity to a great extent, especially in regions marred by water shortage.

Future needs an alternate sources of water - The demand for usable water like any other resource, is increasing fast with the continuous increase in world population and continuous increase in per capital demand linked with more of sewage disposal and transport prolific use of water-using gadgets (like air cooling and conditioning devices, washing machines etc.) and increased recreational activities. According to an estimate, the daily per capital demand for water of an American was 600 gallons in 1900 which increased to 1,500 gallons in 1960 and is expected to be increased upto 2,700 gallons by 2000. The fact is no different in developing world. Thus, the question is where will the water needed for fulfilling increasing demands come from? 'The answer is not so simple and can only be found through long term planning. Possible means of meeting out the impending water deficit problem include: (i) reclamation of sewage and waste water, (ii) development of ground water sources and surface storages;

(iii) long range forecasting of rain; (iv) rain making; (v) transfer of sulfur water, and (vi) desalination of sea water.

Agriculture

The practice through which specific plant species are cared and managed so as to obtain maximum yield of consumable parts of the plant is called agriculture. An agricultural field is, in fact, a cropland ecosystem in which ecological principles involving limiting factors and productivity provide great applicability.

The demand for food production, linked with tremendous population growth, has been increasing fast world over. This has been possible to feed increasing number of mouth through increased agricultural production what is called green revolution. It has resulted chiefly from the industrialization of agriculture, which involves highly domestic plant species, sophisticated chemicals, and large fuel energy subsidics. According to an estimate, to double crop yield through industrialized agriculture, a ten fold increase in fertilizers, pesticides and horse power is required.

The green revolution through industrialized agriculture (fuel powered) as compared to conventional agriculture (mankind domesticated-animal powered) has following backlashes:

(i) Such an agriculture involving huge amounts of fuel, and chemicals in the form of fertilizers, pesticides and weedicides, is the chief cause of environmental pollution.

(ii) It may lead to social backlashes when a quick shift from small farms to large industrialized farms takes place. In this process, small farmers and agriculture workers are driven off the land to cities where they may not get work, food and housing—a situation of despair and social breakdown.

In retrospect, strong measures should be taken to control the population-growth and to improve the conventional agricultural practice with application of current agricultural research under due social and environmental consideration. For a greater and sustained agricultural production following considerations are important:

(i) Available land, acreage should be properly and judiciously utilized.

(ii) Soil fertility should be increased through wise use of fertilizers and organic manures.

(iii) High yielding and disease resistant plant varieties should be introduced.

(iv) Mixed cropping should be practicised wherever possible.

(v) Integrated and balanced use of available water sources, surface and ground water should be made.

(vi) Weeds and pests should be efficiently controlled. Integrated pest control practices should be preferred over total reliance on chemical pesticides.

(vii) Soil erosion and loss of nutrients should be prevented through maintaining vegetation cover throughout the year.

(viii) Exotic varieties should be introduced only after due consideration of long term impacts.

Forestry

A forest is a natural ecosystem having multi-specks and multi-aged trees as dominant community. Forests cover about one-third of the earth's land surface of which about 50 per cent is occupied by tropical forest. Of the total geographical area of India 22.74 per cent is forest as against a minimum of 33 per cent forest cover prescribed under National Forest Policy.

Importance of forests

Functions of a forest may broadly be classified into following three categories:

(i) *Protective functions:* These include the protective role of forests against soil erosion, droughts, floods intense radiation etc.

(ii) *Productive functions:* Forests are the source of wood and many other products like gums, resins, fibers, medicines, katha, honey, bidiwrappers, pulp, paper etc.

(iii) *Accessory functions:* These include the role of forests in recreation, aesthetics, and as habitat of diverge wildlife etc.

Thus, it can be said that forests are important in two ways - ecologically and economically.

Ecological significance: Environmentally, the forests are very important. They help in balancing oxygen and carbon dioxide level in atmosphere, regulate earth's temperature regime and hydrologic cycle. Forests increase local precipitation and water holding capacity of soil, thus preventing drought situation. Vegetation cover provided by forest impedes the velocity of runoff GH soil surface checks soil erosion, silting and landslides, thus reducing the danger of flood. The litter derived from fallen leaves maintains fertility of soil by returning the nutrients. Forests also act as refuge of wild animals and provided protection to them against strong, cold or hot and dry winds, solar radiation, rain and enemies.

Economic significance: Wood which has various applications in domestic and industrial processes, is the chief product of forest. Wood when used as fuel, has certain advantages over coal as its sulphur and ash contents are very low but, at the same times, excessive use of fuel wood gives pressure on forests which have many functions to serve. Wood may also be converted into solid (coke), liquid and gaseous fuels. Timber is an important material in building construction and day-to-day uses. Forest provides raw material for various woodbased industries viz., pulp and paper, composite wood, rayon and other man-made fibers, sport goods, furniture, boat building matches etc. Miscellaneous products, like bamboos, resins, gums, some oils, fibers, medicines, katha, lack, shellac etc., are also obtained from forests.

Depletion of forests

Increased urbanization, industrialization and mining has entailed indiscriminate felling of trees has denudation of forests. Nobody will, however, ever learn that who set the first assault on which tree in the history, but this practice of logging down trees for human use has been going on for centuries unabated. Colonists used logs for fire wood and making huts. They also made handles for their tools, furniture fence posts etc. from it. For them forest had only a present and a past-no future; thus they resorted to ruthless and accelerated exploitation of available forest-a story that still continues with more serious backlashes earth has ever witnessed. India alone is losing more than 1.5 million hectare of forest cover each year and according to Tiwari (1982) 22 million hectare of forests have been destroyed during the last three decades due to over exploitation and misuse and conversion to agricultural fields. The depleted forest wealth would simply deprive the man of economic and environmental values ou'ered by forest, as discussed earlier.

Management of forests

In forestry, the yield is chiefly obtained from the accumulated growth of the past, so this industry has to depend upon the annual growth. Forests are managed to obtain a sustained yield which is obviously possible through sustained growth. Following considerations are important in this direction.

(i) A forest should have in correct proportion the trees of different ages and classes so that continuous removal of exploitable product would not affect the future yield.

(ii) Each tree crop should put on normal growth increments.

(iii) No obstacle should exist to the continuous harvest of exploitable stems.

(iv) Unrestrained felling should be prevented.

For all such planning and decision making, sound ecological knowledge is desirable that include knowledge about climatic conditions of biotic community, successional trend, life cycle of important tree species, methods of arresting growth of undesirable trees and promoting growth of high yield tree species, and biotic inter-relationship (such weeds, competitions, pests, parasites and diseases etc.)

Of various tactics employed in forest management, some deserving special considerations are as follows: (i) restrained felling, (ii) block and selective cutting, (iii) reforestation, (iv) control to disasters, and (v) recycling of forest products.

Rangeland

A range is the grassland that provides natural pasture or grazing animals. Grasslands occur where rainfall is too low to support the forest community but is higher than that which results in desert community. Generally, in grassland precipitation ranges between 25 to 75 cm. However, depending upon edaphic factors, such as a high water table, grasslands may also occur in regions of forest climates. Soil of a grassland is characterized by having high humus content.

Over 6, 000 species of grasses, adapted for a great variety of climatic regimes, cover about 20 per cent of earth's surface. In size, they vary from as small plants as about 1 cm. to over 35 cm. tall bamboos.

Importance of Rangeland

Man has been cultivating various grass species as his food and as fodder for his domesticated livestock since time immemorial. History of human civilization the witness of dependence of man on the seeds of various grasses as his own food, for instance, Indo-Chinese, American and Mediterranean cultures were based on ice, maize and wheat respectively while forage was used as fodder for domesticated livestock.

A range, if not serving as a direct source of food for man, still serves the mankind in a number of indirect ways. Solar energy transformed into food energy by the green cover of the range is available to primary consumers (grazers-domesticated or wild) as food, where in a part of it is converted into meat. Of considerable importance to man are these herbivores and their products such as meat, milk, skins, hides, horns and wool etc. It can thus be said that at the root of all biological resources that, man utilizes is the forage.

Depletion and Degradation of Rangeland

Man would not destroy an artificial pasture developed by input of his hard labour and money than a natural pasture. Thus, like all natural resources the rangeland, inflicted with human stress is depleted and degraded. Cause-of such stress may be ignorance, greed or incompetence of man. Phenomenal growth if cattle industry, on one hand, is responsible for ruthless over grazing. While on the other hand, wide expansion of urbanization, industrialization, and agriculture has resulted in encroachment of rangeland. Often, out of greed too a rancher allows a number of herd to graze beyond the carrying capacity of the range being ignorant to the fact that too many sick and weak herd would not fetch home as much value as a few herd in prime condition of health after grazing on enough good forage.

Community or ecological indicators provide the most practical approach of determining whether or not a zange is being utilized properly (Humphery, 1962), Range managers use the term "decreasers" for palatable species vulnerable to grazing their disappearing is an indicator of grazing stress. When such indicators are not heeded unpalatable "weeds". Once established, it may become hard to eradicate the weeds and not only this such vegetation will invite a variety of pest animals like rodents and insects. This makes clear the long-term consequences of overgrazing in a range.

There is increasing need to restore the depleted and degraded rangeland not only out of economic (livestock production) but also ecological consideration since rangeland depletion would also cause problems like soil erosion, spread of pests etc.

Range management

Stoddard & Smith (1955) deflued the range management as "the science and art of planning and direction range use as to obtain the maximum livestock production consistent with conversion of range resources".

Efforts to manage land include: regulation of grazing pressure: manipulation of stock (herd) distribution: burning of range: reseeding of rangeland: selection of cattle breeds: and control of weeds, pests and predators.

Wildlife

Even on global level there appears to be a great deal of ambiguity attached to the term "wildlife". Webster's Dictionary defined wildlife as living things that are neither human nor domesticated, specially mammals, birds and fish hunted by man. In American usage, term 'wildlife' is used to designate the games and excludes fish, thus they have Fish & Wildlife Services at the federal level. In India too, the common understanding of the people regarding wildlife is of wild animals and shikar, more particularly mammals, birds and reptiles. However, a report entitled 'Wildlife Conservation in India' (1970) of the India Board for Wildlife, states wildlife as the entire native uncultivated flora and fauna of the country. The Wildlife (Protection) Act, 1972, also defines wildlife as any animal, bees, butterflies, Crustacea, firsh and moth, and aquatic and land vegetation which form part of any habitat.

Thus, a comprehensive definition of wildlife may be given as any or all non-cultivated and non-domesticated life including both, plant and animals.

Importance of Wildlife

The Wildlife resources provide aesthetic, recreational and economic benefits and at the same time are important ecologically. Nature's millions of forms of life on this earth perform some or the other definite biological function which keeps the delicate balance of our exosphere. In nature flora and fauna go hand-in-hand and one cannot exist without the other. Bird and insects are necessary for the cross pollination of flowers and these with other animals also serve for dispersal and propagation of vegetation. Not only this, the health of vegetation also depends on natural browsing and grazing. Needless to mention that animals, directly or indirectly depend on green plants for their food. Besides food plant also provide shelter to a variety of animal species equally important are the relations among animal populations. If too many herbivores are killed, the predators, deprived of their natural food, "will leave forest, become vermin, and prey upon domestic livestock or even man. On the other hand if too many carnivores are killed, the herbivores because of release of predation pressure will become too numerous and acquire the status of pest for agricultural crops. Thus most important is the balance of nature of which wildlife in an integral and important part.

Abuse and Depletion of Wildlife

One often comes across the headlines in newspapers like 'Last of the blue whales', save the golden langur Tragedy strikes sea turtles, 'major downfall' in rhino population. Great India bustard on the road of extinction... and so on which narrate the sad story of abuse and depletion of our wildlife heritage. According to an estimate in last 2,000 years the world has lost, by way of extinction about 160 species of mammals and 88 species of birds.

India is endowed with rich biological heritage and her various seripuries and epics are full of vivid and beautiful account on wildlife. Over 500 mammalian and 1,200 avian species contribute to the riches of the wealth. Unfortunately today the Indian wildlife as that in many other parts of the world, is threatened because of man's unprecedented intervention with nature. Nearly 134 plant species have been declared threatened as against 71, 47 and 15 species of mammal, birds, and reptiles respectively (Khoshoo 1984). Great India bustard Choriotis nigriceps confined to the semi-arid regions of Rajasthan and Gujarat, and the white-winded wood dack (*airina scutalata*) inhabiting the Brahamaputra swamps and the flood plains of Assam are both on the birck of extinction.

Wildlife management

The conservation is the practice embracing preservation, maintenance, sustainable utilization, restoration, and enhancement of the natural environment. Wildlife conservation in this sense is the management of nature's living resources if we consider wildlife as consisting both of wild flora and fauna. Since one cannot manage living resources while being indifferent to the non-living environment, the concept of wildlife management becomes the concept of management of natural ecosystems. It also overlaps "forestry" in so far as it is based on multiple land-use concept and implies management of wild plants as well as wild animals. However, under the present dispensation wildlife management implies management of wild animals as distinct from forestry, and it implies maintenance of habitats suitable to the different species.

In America and Europe wild animals, particularly mammals are treated as game crops, as do timber in forestry and food crops in agriculture, and hunting is developed as a viable economic activity. In America, wildlife and wildlife related recreation forms multimillion dollar industry. However, in India it is difficult of develop game as in industry because: (i) our forest areas are already overburdened with grazing pressure of domesticated livestock and because of already scarce pasture lands there is little hope of reducing grazing pressure in the forest areas to develop game: and (ii) hunting as a source of recreation is beyond the means of the majority of population.

During past decades the wildlife management implied the understanding, increasing, and controlling of some species. Sanctuaries and National parks were mostly meant to protect a particular species or a group of species and to limited extent for tourism and recreation. Today, however, the emphasis is shifting from the management of species to the management of ecosystem. The objective of wildlife management, as outlined in the World Conservation Strategy, are thus as given below:

(i) To maintain essential ecological processes and life support systems in natural ecosystems, on which human survival and development depends. These essential ecological processes and life support systems can be defined to include maintenance of both the biotic and abiotic components of the ecosystems.

(ii) To preserve genetic diversity (i.e., the range and variety of genetic material found in the world's wild plants and animals) on which depend the breeding programmes necessary for the protection and improvement of cultivated plants and domesticated animals, as well as much scientific advance, technical innovation, and the security of the many industries, that use living resources.

(iii) To ensure the sustainable utilization of species and ecosystems. This will include the proper utilization of wild animals for game, forest and grazing, and to achieve proper integration for economy both of rural communities as well as trade and industry.

Fortunately, people have been alerted now and the strong measure, being taken in favour of restoring, maintaining, and increasing wildlife population given hope that our rich legacy of wild life will not only be saved for future but developed further. Ways and means employed in wildlife conservation include: (i) protection of breeding stock by enacting game laws and developing wildlife refuges; (ii) artificial stocking including introduction of exotics; (iii) control of predators; (iv) habitat improvement and (v) garhe farming.

Minerals and Their Characteristics

Minerals are the non-living substances obtained from the earth through the process of mining. Formation of minerals took place predominantly in weak zones like mountains, folded and faulted regions, lakes, troughs, and continental shelves through geological processes going on since billions of years. They hold some peculiar characteristic, as discussed below, which distinguish them from other natural resources.

(i) Exhaustibility

Minerals are non-renewable or depleting assets and once mined out the deposit get depleted and lost for ever without any chance of replenishment. Refuse or waste, however, can be consumed afterwards through improved benefication technology.

Simply this exhaustible resource cannot be cropped year after year like agricultural products. Therefore, three distinct stages can be identified in the life cycle of a mine: youth-developmental stage; maturity established stage; and old age-declining ore reserves. If the rate of mineral consumption is increasing fast, the knowledge about more reserves and potential substitutability of many minerals is also being improved. As such these predictions may prove outlived.

(ii) Localised occurrence

Minerals are very localised in their occurrence and the total area occupied them is far limited. Most of the mineral deposits are very sporadic in distribution, for instance, Kolar gold fields of India occupy an area of 8 km. by 1.6 km. formations containing coal and petroleum are; In general, bigger as compared to metalliferous deposits. The coal-bearing formation of India, for example, occupy about 4 per cent of the country's total area.

(iii) Unpredictable occurrence

The quantity and quality of a mineral in any region is not easy to predict. Uncertainties prevail with respect to their occurrence, extent of depth, cost of mining (often linked with depth) and the grade.

(iv) Risk involvement

The process of mining involves large number of problems for miners. Abnormal temperature and pressure variations, rock burst, fire and gas hazard, lack of ventilation and drainage, collapse of roof and side walls are the peculiar problems encountered only in mining industry.

(v) Dependence on other factors

Mineral production is largely affected by factors like the changing needs of consumer industries, availability of stock, fluctuation in prices, and reuse of metallic scrap in production process. A technologically poor country possessing richer grades of ores may have to face keen competition in international market with technologically richer countries may, however, possessing poorer grades or ores since in later case the cost of production is considerably low due to improved mining methods.

(vi) International character

Minerals know no political boundaries. Virtually, no country within its own boundaries has adequate production and reserves of all the minerals needed for its industrial development. The unequal distribution of minerals makes the international trade in its inevitable. Some countries are bestowed with monopolistic rights in respect of certain mineral deposits which result in what is known as geopolitics.

Development and conservation of Minerals

Geologists are of general-apprehension that the known reserves of minerals may not last for long and most of them will be exhausted within 100 to 200 years. New possibilities of mineral reserves in land and beyond inland (as in the sea-bed), are being explored which may reduce, if not totally remove, the futuristic fears of exhaustion of the resource.

Presently, ilmenite, rutile, tin ore, diamond, sulphur, coal petroleum, and calcareous and ferruginous sea sands are largely being mined from the seacoasts and sea beds. In India, ilmenite, rutile and associated zircon, monazite, and sillimanite are obtained from the beach sands of Madras and Kerala coasts. Coral limestone for industrial use like manufacture of calcium carbide, and calcareous sand for cement are being mined from the sea-bed at Tamil Nadu and Dwarka (Gujarat) respectively for quite long time. Indonesia, Thailand, and Burma recover considerable tin ore from the sea-bed, while

England recovers coal and Philippines iron from the sea-bed. A number of petroliferous structure occurring off the coast are being mined for petroleum and its products at several places. The ferromanganese nodules or what commonly known as manganese nodules found over the oceanic floors, generally below 4,000 m depth, are perhaps the most spectacular discovery in the field. The most prolific concentrations of such nodules with high metal values have been found in North Pacific between 139 W to 141 W and 4 N. Although precise information about the reserves of manganese nodules is not available but it is estimated that they may run several billion tons and may prove to be potent source of metals like manganese, nickel, copper and cobalt. It is estimated that 1 million tons of manganese nodules from sea-bed provide about 275,000 tons of manganese, 13,000 tons of nickel, 2,5000 tons of cobalt, and 1,000 tons of copper besides other metals like iron, vanadium, molybdenum, silver and lead. Further, it has now been roughly estimated that the metal resources available in manganese nodules are several times greater than those available in land, such as cobalt 5,000 times, manganese 4,000 times, nickel 1,500 times and copper 150 times.

The conservation of mineral resources does not mean preservation of minerals under ground to be used by future generations but it implies efforts to utilize minerals in best possible way, depending upon the industrial needs and changing technological requirements, so that the life of the deposits is increased without sacrificing out own needs. From the utilization point of view, minerals are placed under two classes: expendable and non-expendable. Expendable minerals have practically no salvage value and ones consumed are lost for ever, e.g., coal, petroleum, gas and most of the non-metallic minerals. Non-expendable minerals are those which have good salvage value and can be used again and again through recycling, except those quantities that are lost due to corrosion and abrasion, e.g., metallic minerals.

Efforts to conserve mineral resources include:

(i) Application of efficient methods of mining to take out every possible tonnage lying under ground.

(ii) Utilization of unmarketable ores through the innovative methods or ore dressing and metallurgical practices.

(iii) Use and reuse of scraps.

(iv) Recovering all associated elements as coproduce or by-product.

(v) Economic use of minerals.

(vi) Substituting the use or rare land costly minerals with those which are abundant and cheaper.

A facility (house, hospital, port) or even in providing a service to society (medical care, public transport). Firewood, fossil fuels (coal, petroleum and its products, natural gas), solar energy, biogas, wind and hydro power, and atomic power are chief energy resources. Of these, firewood and fossil fuel are being used for supply of energy since time immemorial and are known as conventional energy sources. Wind and hydro power, solar energy, biogas, and atomic energy are the new and non-conventional sources of energy. Fossil fuels being limited in supply may be termed as exhaustible sources while others, like solar, hydel, wind and atomic energy, that would be available to man for infinite period are regarded as inexhaustible sources. Firewood and other biofuels (Such as, biogas derived from biological wastes, domestic garbage etc.) are indeed renewable energy sources but exploitation of fire woods has great ecological implications since we are already short of forests and whatever forests available should not be squeezed or thinned. However, use of other biofuels is known to be the best option in the light of two important acts: their steady of even increasing supply at low cost; their use in energy production reduces pollution of environment.

Because of rapid depletion and poor regeneration of the exhaustible sources of energy the society should concentrate on tapping the inexhaustible and renewable sources under strict economic and ecological consideration.

Conventional and exhaustible energy sources

Fire wood: No one can trace back in the history that who has made the first assault that is on which tree. However, it is evident that man has been logging down the trees for various purposes, including to get firewood as an important one, since long. Because, in olden days man's knowledge and access for various natural resources were limited, he ruthlessly exploited his own surroundings to meet out his day-to-day requirements. The story still continues in various parts of the world at sic. ui or faster pace, but often not because of the dearth of knowledge but out of vested interests or excuse of having no other option. Felling of trees for firewood is still a common practice in interior where it is the cheapest and, at times the only available source of energy. This had led to thinning or denudation of woodland that needs serious consideration from both ecological and economic viewpoints. Obviously, the only wayout to the problem is making available to the villager the cheaper non-conventional energy sources, such as biogas and solar energy in usable form.

Fossil fuel: Fossil fuels (coal, petroleum, natural gas) are the most extensively used sources of energy today. Like for many other natural resources increase in population and per capital energy demand coupled with the industrialization at a faster pace, are the factors responsible for depletion of fossil fuel world over. According to a report the earth's original stocks of coal and petroleum were about 6,000 billion tons and 1,250 billion barrels respectively, of which by the year 1960,100 billion tons of coal was mined and 100 billion barrels petroleum extracted. The story of natural gas is no different which has acquired the status of irreplaceable commodity of domestic use in modern times. Petroleum and natural gas today fulfil 60 per cent of the world's total energy requirements. It is evident that if the exploitation and consumption of these resources continues at the present incredible rate, there supply may last only for a few more decades.

Non-conventional and inexhaustible energy sources

The rapid depletion conventional energy sources has prompted governments and people to concentrate on finding and tapping some non-conventional energy sources that may last for long. In fact, inexhaustible energy sources, like solar, hydel, tidal, wind and atomic power, can only bring hope for the sustained socio-economic development of humanity for years to come when exhaustible sources will become either inaccessible or too costly to afford.

Solar energy: Research and application in the field of solar energy have opened new vistas in the direction of fulfilling world's future energy requirements. It has especially drawn attention for its practically free steady supply and pollution free use. This energy may be transformed into heat using reflectors and condensers, or into electric energy using solar photovoltaics. The resource can especially be tapped more effectively in the regions where there are long sun-shine hours.

Solar heaters and cookers are becoming popular these days in industry and household. Passive solar architectures provide comfortable housing in extreme cold and hot climates. This engineering has also been successfully applied in warehouse and cold storages. A multipurpose solar dryer-cum-warehouse is under construction at Ganaur (Haryana, India) where chillies, potato-chips and other foodstuffs could be dried and stored. Solar energy has also been effectively used in desalination of water. College of Engineering, Pune has designed a solar furnace that can generate heat upto 2,000^{o} Cu which can be send in various metallurgical and other purposes.

Department of Non-conventional Energy Sources of Govt. of India has prepared a perspective plan that envisages generation of energy through non-conventional sources, chiefly solar to the extent of 250 million tons of coal replacement per annum including 15,000 MW of electricity by the year 2001. Indian's first solar power station, soon to be operated near

New Delhi, would be a great breakthrough in this direction. Rural electrification through solar photovoltaics is already in operation at some places in the country. Moreover, photovoltaic irrigation pump sets have also been installed at some places.

Wind power: The power of wind since olden days is being used in running mills, irrigation of fields, and carrying out other farm activities, but its real potential is not so far being realized. According to inventories of Department of Non-conventional Energy Source of Government of India 20,000 MW electricity can be generated from the wind alone in the country. Wind power plants are set up in various parts of the country but they still account for the total capacity of 3.3 MW only. Obviously, this potent resource is still to be tapped optimally.

Hydro Power: Use of gravitational flow of water in generation of electricity, what commonly known as hydro power or hydel power is quite common in various parts of the world. India's, first hydro power station, having a potential of 300 KW, was constructed in 1897 at Darjeeling. Later on, many stations with far high potentiality were set up. The country's economically exploitable hydro power resources have been estimated at 41 million KW (or 41,000 MW) of which only 18 per cent of the potential was utilized by 1979. The Brahmaputra basin has the highest hydro power potential and nearly 30 per cent of the country's total; next to it rank the Indus, Godavari, and Ganga basins respectively.

In world, a total 4,910 × 100KW electric power is generated of which hydropower forms of fourth. In highly industrialized countries like USA, USSR, Canada, Japan, France, Sweden, and Norway hydro power is nearly fully developed whereas in the rest of the world it is yet to be developed fully.

Hydel power is considered to be the cheapest source of electricity with the cost of generation ranging from 2 to 3 paise per unit, compared to a minimum of 5 paise per unit of-electricity generated by coal based thermal power plants.

Although the initial cost of hydel project is generally 25 per cent higher but owing to lower recurring cost, the cost of generation of hydropower is the lowest. A part from such economic considerations, development of such projects is beneficial for irrigation and other purposes too.

Tidal power of water in costal areas of India, however not so far being realized. According to an estimate a tidal power potential of 9,000 MW has been identified of which 7,000 to 8,000 MW is in the Gulf of Cambay, 800 to 1,000 MW in the Gulf of Kutch and rest in sunderbans.

Atomic power (Nuclear power): As already delineated, the fossil fuels beings exhaustible, such sources of energy are not to last for long with the current trend of consumption. Atomic power appears to be the only hope for large scale energy requirements when fossil fuels are exhausted. The reserves of atomic fuels are said to be over 10 times greater than fossil fuels and the advantage of former is that their very small amounts would produce enormous amount of energy. A ton of Uranium^{235} for instance, would provide as much energy as 3 million tons of coal or 12 million barrels of oil. Atomic energy has its application not only in generation of electricity but has successfully been used in marine vessels, spaceships and even in chemical and food processing industry.

Important constraints in atomic energy generation are: (i) cost of construction and maintenance of plants; and (ii) disposal of radioactive wastes cost of construction of an atomic power plant and its maintenance is very high as compared to any other electric power plant. For running one atomic power plant unit of 235 MW capacity, heavy water of a worth Rs. 10 crore is required per year. Further, as a consequence of frequent failures, the running cost of such plants come to be very high.

Moreover, after the working capacity of an atomic power plant is over (in about 25 years), these have to be destroyed so as to avoid radiation spillage. Surprisingly the cost and time for destruction of these plants is almost equal to that for construction. Disposal of radioactive wastes is such a problem that it until has no satisfactory solution for long term.

Bioenergy (Energy from biological matter): Organic wastes provide an important renewable source of energy. Biogas generated from the decomposition of organic wastes, like manure and domestic wastes is becoming increasingly popular as a source of heat and light in many parts. It is considered to be advantageous in view of its relatively cheaper supply, and use of organic wastes in its generation reduces the otherwise impending threat of pollution due to their release in environment. As such, it serves two purposes; fuel production and waste disposal. It has more of practical applicability in villages where organic waste in the form of cattle dung, is available in plenty. Biogas so generated is thus economic and convenient to use as compared to conventional practice of burning of dry dung-cakes. Further, the dung-cakes serve only as a source of heat for cooking and allied purposes while biogas has multifarious applications. The slurry so produces in a biogas plant can well be used as manure for agricultural fields.

Pic. 2.1: Jungle Get-Away : The lake at Tadoba is an ideal place to spot the big cat.

Pic. 2.2: Fishermen repairing their boat in Campbell Bay.

Pic. 2.3: The park holds miles of trails that provide vistas of the spectacular landscape.

Pic. 2.4: After 200 million years : At the kew Gardens in the U.K. where it was planted recently.

3

COMMUNITY ECOLOGY AND POLLUTION

When rainwater falls on an uneven surface, it collects in depressions. As the water overflows them, the current erodes a narrow channel that deepens with each succeeding shower and may eventually drain the depression. There is usually also a lateral meandering of the stream by which a valley is formed. The site of the waters of such streams is impermanent and continued erosion forces the waters and the channel farther and farther back into the upland. The stream is at first a temporary one, dependent for its waterflow. On rainfall runoff but when its channel is cut below the level of the groundwater table the stream becomes permanent fed by general continuous seepage. The waters of such stream are therefore its youngest portion physiographically and the stream is progressively more aged towards its mouth.

In hilly or mountainous terrain, water may accumulate in large basins until ponds or lakes are formed. Ordinarily, however, the water level in a lake will rise until it overflows at the lowest point on the perimeter. Then the waters continue to flow downwards until they reach the sea. Streams springing from fixed headwaters (melting snowfields and glaciers springs) carve valleys throughout that are essential, the same age.

A river system in youth is characterized by valleys that are narrow and steepsided; the flow of water is usually fast, there are few tributaries and there are many waterfalls, ponds and lakes along its course. As the river system matures, its valleys become wider, its slopes more gentle, and its tributaries more numerous and longer. Many ponds and lakes are drained, and waterfalls are worm down to rapids or riffles. The areas of upland are well dissected and the lands are thoroughly drained. In old age, the river stream has reached base level.

The upland has been worn down to low ridges between tributary river valleys and the regions as whole is called peneplain. There are no lakes ponds or rapids and the flow of water is sluggish. Strahler streams less than 3 m wide are called Creeks or Brooks. Rivers are streams 3 m or more wide.

Habitat

Exclusive of its lakes, the principal habitats in a stream are falls, rapids or riffles or sand bottom pools and mud-bottom ponds. The character of the bottom depends primarily on the velocity of the water current, which along with the volume of the stream flow can be easily measured. Water flowing at the rate of about 50 cm-sea is considered swift flowing velocity greater than 300 cm. sea rarely occur. Fast currents rolls or slide peddles and rocks along the bottom, move sand partly buoyant transportation, and carry fine materials, such as silt and organic matter, in suspension.

In place where the topographic gradient is steep the bottom will be composed largely of cobble and boulders too heavy to move, and smaller pebbles which are trapped by obstructions. This habitat is called a rapids, if extensive and turbulent riffles, if of lesser order. Below the riffles there may be a raceway, which is intermediate between the riffle and pooi in depth of water, rate of flow, and substrate and its depth is quite uniform.

When the gradient is less steep and the water current thus slower gravel (particle size 2-64 nm) is deposited first, than sand (0.06-mm) but the finer materials are carried along. Only when the current becomes negligible downs the suspended materials settle so that silt (0.004-0.062 mm) or mud-bottom pools and ponds are formed. Clay has a particle size even smaller. These mudbottom pools are fertile parts of a flowing, non-polluted stream. In a general way, riffles and sand and mud-bottom pools represent three stages in erosion cycle of a stream, and ecological study of them gives a good idea of what the erosion would be over a long period of time.

STREAM BIOCIES

When quantitative sampling is made of the invertebrate populations of streams one finds that there is a sharp distinction of species found in riffles and those found in mud-bottom pools. The raceway is here included with the riffles. The sand-bottom pool habitat has no characteristic indigenous invertebrate species, but is occupied by small numbers of individuals of species otherwise occurring abundantly in the other habitats. The unstable bottom apparently prevents the development of distinctive community. The high index of species diversity is caused by the uniformly low number of individuals in each species which more man compensates for the smaller number of species present. There are however, several fish species that find sandy pools a favourite habitat although they depend in large part upon riffles organisms for their food. Stream fishes commonly segregate into more or less distinctive societies of species correlated with differences in current type bottom and depth of water. Many fish over-winter in deeper, more quiescent sand bottom pools especially since low water temperature makes them sluggish to withstand rapid currents.

The stream biocies consist most typically, therefore, of the inhabitant of the riffles and sand-bottom pools found throughout the course of the river. The riffle and pool organisms make up two different fades in this community. These facies become more fully developed downstream as shown by an increase in the species diversity index, both for benthic organisms and for fish. Mud-bottom pools and sluggish streams are occupied by the pondmarsh biocies.

Plants are not abundant in the stream biocies, although the upper surfaces of rocks in a riffle may be completely covered with branched filament algae (particularly Cladophora) and a few species of water moss (Fontinalaceae) may occur. Diatoms mostly sessile form, may be numerous in early spring and again in the autumn. Dominance in the true sense such as occurs in terrestrial communities, does not exist, although the algae and mosses passively provide food and shelter for active forms.

The most characteristic and abundant animal forms of the stream biocies are the caddisfly larvae, mayfly naiads, stonefly naiads, fly larvae, crayfish, snails and calms sponges, and bryozons, and fish each occupying its own particular niche plankton is mostly sparse in swift-running water, but may be abundant in sluggish, pond like stretches of large rivers.

Adjustments To Current

Probably the characteristic of stream is most critical to the life therein the current. All organisms that occurs in streams must adjust to maintain constant position. Torrential floods occur the stream bed, move rock and sand, cut new channels, and destroys entire populations. Recovery after such catastrophe may take place within a few weeks, especially by those species possessing short life cycles but sometimes it may require years (Moffett 1936). Position is ordinarily maintained by clinging to the substratum avoidance of the current or vigorous swimming, and requires a good development or orientation behaviour.

Clinging Mechanism

The growth from fresh-water sponges is affected by a number of factors, but in riffles sponges are usually simple encrustations. In quieter water long slender finger like processes may form. The distribution of species depends both the current and organic content of water.

Plumatella is common bryozoan that forms an encrusting plantlike, branching colony on the underside of rocks or fallen trees in swift water. Pectinatella in the other hand forms a gelatinous spherical ball, and is more commonly found in ponds or slow flowing portion of streams.

Respiration And Oxygen Requirements

Oxygen is usually ample in streams, often saturating the water in turbulent riffles. The oxygen concentration is sometime low however, in sluggish streams and standing pools. The differences in oxygen tension of the two habitats is reflected in the respiratory adaptations of the organisms that inhabit them.

The lemelliform gills of the mayfly naiads inhabiting mud-bottom pools are larger in size than those of species and are almost continuously flicked back and forth for better aeration. The gills reduced by two-third in proportion to body weight compared with mud dwelling forms. They are never flicked since the water movement continually brings oxygen to them.

The respirator equipment of pond-inhabiting animals permits them not only to live in habitat with lower oxygen tensions but also to survive longer at high water temperatures. Often, these animals display relatively low rates of general body metabolism and oxygen requirement. Such relations between riffle and pond animals have been observed for mayfly naiads, caddisfly larvae isopods, crayfish and fishes.

Preservation Of The Environment

The chief problems in applied ecology related to streams and other aquatic bodies are those of maintenance of water supply for man, flood control recreation, fish management and control of pollution. Pollution takes four forms silting resulting from excessive erosion of the surrounding upland, industrial products produce by inorganic chemicals from water lead and zinc works, tanneries, breweries, paper mills, gas plants, mines etc. thermals hot water effluents from atomic plants and organic principally municipal sewage, from water from agricultural land.

Erosion and Silting

Stream erosion becomes considerable when upland vegetation is to be reduce that there is little or no retardation of runoff from heavy rains. Continuous erosion throws a heavy load of fine silt into the stream. This is detrimental. It makes the water opaque reduces or prevents photosynthesis in algae, water moss, and other plant life handicaps those fish and other animals on sight finding and capturing food and clogs the filtering mechanisms of various invertebrates. Deposition of silt on rock or sand bottom may bring a considerable change in species composition of animals present. During the last several decades greatly increase soil erosion on agricultural areas has reduced pan and game fishes in the streams.

Chronically muddy streams may often be cleared by reforesting the waters head and by practising modern erosion control in cultivated areas. With slower runoff more rainwater soaks into the ground and the water is raised. It is also desirable to maintain vegetation on the immediate stream banks to slow up under cutting.

Stream band vegetation is also beneficial for shading the water and keeping it cool enough for such fish as trout. If artificial dams are necessary they should be small and located where the drainage brings in the numerous head water of the streams.

Water Supply, Food Control, and Recreation

As the human population has expanded during the last century, man has exerted more and more control over the stream flow to conserve it for drinking and home use, irrigation of lands, industrial and municipal use, navigation, swimming, and boating, to produce electric power and to prevent flood damage. This extensive use commonly involves construction of dams and reservoirs, dredging of river channels raising of levels along the shores to confine the course of water flow rates. As a result, the riffle habitats are eliminated, water levels are altered, and flood plains are permanently inundated. The stream biota disappears to be replaced by mat of lake or pond, mudflats are created with drawdowns or utilization of water in excess of the inflow, and consequently the environment is drastically changed. Unfortunately, sedimentation and filling of the basin with silt eroded from the upland often limit the life of such reservoirs to 50 to 100 years. There is a limit in height to which a dam can be repeatedly raised and to the number of reservoirs that can be built along the main channel of river. Alternative solutions are tapping of underground aquifers for water supply, construction of multiple dams on headwaters of tributary streams for flood control and restricting water use to what can be rapid between of the ground water reserves.

Industrial Pollution

Industrial and mine wastes are often acid, extreme acidity will kill fish and other organisms. Clams are greatly reduced to disappear altogether in acid waters. Industrial wastes contains a great variety of chemical compounds, including salts of the heavy metals and many of them are toxic to fish. Young fish as species of small fish appear especially sensitive, and the pollution materials may cause physical or chemical injury to the gills without actually being absorbed into the body.

Mercury pollution began to be a problem in Sweden and Japan in the mid-1950's and has since then become of worldwide concern. Trace amounts of mercury naturally present in environment but the difference between tolerable natural levels and levels harmful to man and animals is very small. Mercury is used in agriculture and paper mills, as fungicide and this treatment have caused mortality in terrestrial birds. Mercury from industry and from combustion of some coal and petroleum becomes washed into fresh-water streams and lakes and eventually into coastal water of ocean. Metallic or inorganic mercury does some damage to animals, but ordinarily it does not accumulate in the body to serious levels. When acted upon by micro-organisms, however, methylation occurs and methyl mercury may become so concentrated, via the food chain in fresh and saltwater fishes as to be dangerous to man and fish eating birds.

The control of radioactive wastes from uranium mills and other atomic energy plants and from fallout has become an especially serious problem times. No stream or other aquatic body can purify itself of these waters through biological means. However, they become dilated downstream, undergo physical decay settle out in the bottom and are taken up by the organisms. Organisms assimilate elements at equal rates whether the elements are radioactive or not. Some radioactive elements have a lift-life extending over several years and may thus accumulate and become concentrated in organisms to an extent many thousands times greater than their concentration in water. This is of potential harm to man when he consumes these organisms. Mammals are most sensitive to radioactive materials, than, in order, seed plants and lower vertebrates, insects, and micro-organisms.

Thermal Pollution

Water is required in atomic reactors for absorbing the waste heat that is generated, and heated water is then returned to the stream lake or ocean. Engineering devices are available to recool the effluent and either recirculate it through the reactor. Return it to the stream, or use it for other purposes, but to do this has been considered uneconomic. A challenge to engineering is to use the heat generated for useful purpose, such as heating homes, in irrigation of crops, in industry, and so on.

Some bacteria blue-green algae, rotifers nematodes, and protozoans can survive in water up 50°C to 60°C, but very few higher organisms can tolerate aquatic temperatures above 35° to 40°C. Infact most animals come under stress at significantly lower water temperatures. With the rise in water temperature, the solubility of oxygen decreases whereas that of salt increases. Various physiochemical changes occur in the functioning of cellular membranes, hydration of enzymes, and so on. Chemical reaction such as those involved in decay of organic matter and other oxidation processes proceed more rapidly in warm water than in cool water. Likewise, metabolism of organism and hence their demand for oxygen increases in warm weather. The augmentation of metabolism and oxygen requirements of organisms may well exceed the capacity of photosynthesis for producing oxygen. The resulting change in habitat brings a drastic change in the species composition of the biota or, when extreme, its elimination.

The heat produced by atomic reactors, as well as heat generated by man in industry and modern living generally, eventually reaches the atmosphere. The total heat now being produced by man is estimated at 1 per cent of the total radiated by the earth. This is too small to have a significant effect on the average temperature of the earth's surface. However, man's output of heat is increasing about 7 per cent per year.

Organic Pollution

The introduction of small quantities of organic wastes may increase the size and productivity of plant and animal populations by adding to the basic nitrogen and phosphorus supply (Flemer 1970). The limit of the sewage load that a stream can carry without harm is however, low and soon reached. As fresh organic material oxidizes, carbon dioxide and toxic gases are released into the stream, and there is a drastic reduction in the oxygen content. Fermentation is more rapid in summer in water, and may begin in wastes before they are discharged into the stream. The decomposing organic material continues to be oxidized through bacterial and other action as it is carried down the stream, and when this action is completed the stream is again pure.

There have been many attempt by means of chemical analysis of the water to determine the degree to which stream is polluted and the rapidity at which it purifies itself. There is difficult, however, in evaluating the extent to which each of the many chemical compounds to be found is harmful to the various organisms. There is considerable variation in this respect, even between different stages in life cycles of the same species. Furthermore, the sewage load may vary from time to time, and infrequent heavy loads may wipe out the animal life in localities where chemical measurements made at other times do not indicate harmful pollution.

Various investigators have attempted to use invertebrate animals as indicators of pollution. The presence of midge fly larvae, Tendipe sriparious, Glytotendipes, mosquito larva Cluex pipiens, Pattail maggot, and sludge fly delimit zones of septic pollution. There are relatively few species that can tolerate septic conditions, but those that do may become very abundant.

Lakes And Eutrophication

Lakes are large bodies of fresh water usually deep enough to have a pronounced thermal stratification for part of the year.

Lakes are formed in youthful stages of river system development. Water from upland runoff, groundwater seepage, and melting snow fields and glaciers collects in basins. As the basins overflow, erosion of outlets continues and the water level of the lake drops. Products of erosion from the surrounding upland are carried into the basin by wind and water, and along with the products of animals and plant decay fill in the basin in the course of time.

Morphometry aside the essential distinction between lake and stream habits is the characteristic of water movement continuous rapid flow is characteristic of the stream the lotic habitat. The lake is a lentic habitat, the water is essentially a standing quiescent body, although at times wind action stirs surface layer and margins into great turbulence.

Habitat

The aquatic habitat offers a variety of unique conditions to which organisms must be adapted before they can occur.

Pressure, Density and Buoyancy

The pressure imposed on a lake dwelling organisms is the weight of the column of water above it plus the weight of the atmosphere. Most lakes have maximum depth of less than 30 ml. Maximum pressure are much less than in ocean, and organisms appear to adjust them readily. The absence of animal life in deep water is ordinarily a consequence of low oxygen supply, or low temperature rather than pressure.

The density of water varies inversely with temperature and directly with the concentration of dissolved substances. Water is most dense at approximately 4°C and becomes progressively less dense as it is cooled below +4°C. Ice also extends markedly (that is, becomes less dense) the colder its gets. It is because the coldest water is at the surface in winter than ice forms there, rather than at the bottom. In summer, the coldest waters of deep lakes are at the bottom. Dissolved salts increase the density of water, the density of most inland water bodies is much less than that of the ocean.

By the law of Archimedes, the buoyancy of an object is equal to the weigh of the water displaces. Buoyancy varies with density of water and is influenced by the factors that affect density, Viscosity, the measure of the internal friction of water, varies inversely with temperature and also influences buoyancy.

An organism will sink unless. It keeps by swimming movements, or unless it has special adaptation to decrease the specific gravity of the body and take advantage of turbulence in water. Such adaptations take several forms, absorption of large amounts of water to form jelly like tissue storages of gas or air bubbles within the body, formation of lightweight, fat deposits within the body, or oil droplets within the cell increase of surface area in proportion to body mass, which increases frictional resistance. When an organisms so equipped dies the special mechanisms quickly, cease to function, and it sinks to the bottom. If dead organisms did not sink to the bottom, living organisms, with the exception of some bacteria, could not exist in an aquatic habitat.

Light

The daily alternation of light and darkness establishes a rhythm in the activities of many aquatic organisms. Light is essential for photosynthesis, some fish require light for feeding. Many organisms orient to light and some are sensitive to light of particular wavelength notably ultraviolet. Small, soft bodied, bottom dwelling organisms are particularly sensitive to light, and it is though that the evolution of pigmentation, chitinous exoskeletons, sheels, cases and similar structures may have helped certain otherwise photosensitive species to survive in shallow, well lighted areas.

Suspended material in water produces turbidity, and reduces light penetration. In western lake Erie, the depth to which 1.0 per cent of surface light penetrates varies from 9./ m, when turbidity is 5 ppm, to 0.8 in. When turbidity is 115 ppm. Since photoplankton requires light for photosynthesis, abundance varies inversely with turbidity. Light penetration is also affected by the abundance of organisms themselves, both by phyto and zooplankton.

An appreciable amount of light passes through ice in water. This enables phytoplankton photosynthesis to continues. In eutrophic lakes many fish suffocate when snow overlies surface ice, preventing photosynthesis.

Photosynthesis releases oxygen into the water, respiration and decomposition absorb it. The upper layer of a lake, where photosynthesis predominates, is called the trophogenic zone. Below this zone there may still be considered photosynthesis, but oxygen absorption is greater than oxygen release. The deeper portion of a lake is called tropholytic zone. The two zones are separated by a thin layer where the oxygen gains photosynthesis during the daylight is balanced by the respiratory and decomposition losses during the day and night. This is the compensation depth, to which generally about 1 per cent of the full sunlight at the water surface penetrates. The compensation level in a darkcoloured bog may lie less than 1 m below the surface, in a deep clear lake it may be 100 m down.

Wind and Currents

Wind is an important environmental factor of lakes because of the water currents it generates. The effect of wind action depends largely on the exposed water surface, the presence or absence of protecting upland, and the configuration of the lake relative to the prevailing wind direction.

Waves may become sizeable in large lakes, but the forward motion of a wave does not involve any great mass water. The rate of movement of surface water is usually less than 5 per cent of the velocity of the wind.

The water washes back of the beach as an undertow, only to be carried forward again by the incoming waves. The problem of maintaining position here is similar to the problems of maintaining position is streams. The depth of wave action in the open lake and along the shore depends largely on the strength of the wind.

In summer, surface water is warmed by solar radiation and its density, weight and viscosity decrease. In deep lakes, the warm water piles up on the exposed shore until moving down along the bottom, it encounters colder and denser waters which resist mixing. The warm water is then diverted horizontally to the opposite shore. Thus, the lake becomes stratified horizontally into an upper epilimnion, where the water circulates and is fairly turbulent, and a lower hypolimnion which is relatively undisturbed. This difference in circulation in deep lakes is closely correlated with differences in temperature and oxygen characteristics, it is of considerable importance in the distribution of the biota.

Temperature

The thermal conductivity of water is very low, but because of the thorough mixing of the waters in the epilimion during the summer by wind action, the temperature is nearly uniform down to the thermocline. The thermocline is the zone of most rapid temperature decrease, generally involving a drop of at least 1°C per metre depth and occasionally as much as 7°C/m. The thermocline, as here defined equals the metalimnion of some authorities, who have a different conception of the thermocline sinks. When the thermocline forms early in the season it is close to the surface. As the season progresses, it sinks lower increasing the volume of the epilimnion and decreasing the volume of hypolimnion. During autumn, the surface water cools and thermocline sinks. The epilimnion increases in thickness until it includes the entire lake. The waters are then uniform in temperature and density at tall depths. This is the autumn overturn which may last for several weeks or until ice forms. As ice melt during spring and surface of water warms up, a spring overturns occurs when water at all depth is at the same temperature when a lake has to overturn during the year, it is called dimatic lake.

Oxygen

The distribution of oxygen at various depths depends upon the presence or absence of thermocline, the amount of vegetation and the organic nature of the bottom. The amount of oxygen in water is only one-forth to one-twentieth of that present in equal volume of air when the two are at equilibrium, although their partial pressures are the same. Diffusion of oxygen from the air into comparatively sedentary water occurs very slowly, agitation of the water increases the surface areas and promotes a faster rate of equilibration.

The oxygen supply of lakes is reduced in various ways, most notably through the respiration of animals and plant and the decomposition of organic matter. As lake water warms up during the summer, their capacity to hold oxygen may be released into the atmosphere. The saturation capacity of water at 0°C is 102 cc/litre, but at 25°C it is only 5.8 cc/litre. In some lakes, decomposition of organic material at the bottom may deplete the hypolimnion of its oxygen content for several weeks during the summer lower than the level minimal to the support of many forms of life. This is called the summer stagnation period. During the winter, if the lake is covered with ice and snow, there may be a winter stagnation period. The oxygen supply of the water is renewed with the autumn and spring overturns. Before decomposition can proceed very far there must be calcium in water. Hence, decomposition is slow in soft or acid waters.

At temperature of 15-26°C oxygen concentration of less than 2.4 cc/litre (3.5 ppm) are fatal within 24 hours to several species of fish. From 3°C-4°C, oxygen concentration can decline through 48 hours to 1.4 cc/litre (20 ppm), or even to 0.7 cc/litre (1.0 ppm), before the same mortality results. This is related to the reduced metabolic requirements of cold-blooded organisms at low temperatures. Some planktonio-invertebrates can tolerate oxygen concentration as low as 0.2 cc/litre (0.3 ppm). Some bottom dwelling protozoans, annelids, mollusks, and insect's larvae may survive in actual anaerobic conditions for periods of days, even weeks. Organisms that tolerate a lack of oxygen do so by creating an oxygen debt, that is the lactic acid and other breakdown products produced in consequence of muscular activity simply accumulate until conditions permit oxidation of them.

Carbon Dioxide and Other Gases

Carbon Dioxide is required by plants for photosynthesis. Its presence in lake waters tends to vary inversely with oxygen.

Carbon dioxide is derived from atmosphere, the respiration of both animals and plants, decaying organic matter, groundwater, and bicarbonates (dissolved CO_2), half bound state (HCO_3) or fixed state (CO_3). These states are associated respectively with pH value 7, 7 to 10 and above 10. Algae and some rooted aquatic vegetation are able to obtain the half-bound CO_2 from the soluble bicarbonate salts, thereby converting them into less soluble carbonates.

A few insects and some bacteria are also able to precipitate carbonates. Methane hydrogen sulphide, nitrogen and ammonia are other gases present in a lake which affects the biotic environment.

Lake Biocies

Depending largely on their morphological adaptation and behaviour, aquatic organisms are for convenience, divided into plankton neuston and benthos although the differences between the groups are not precise.

Seston is a collective term that includes all small particulate matter both living and non-living, that floats or swims in the water. Plankton are free floating or barely motile organisms, either plant (phytoplantkon) or animal (Zooplankton) that are readily transported by water currents. Most plankton are microscopically small although some forms are visible to the unaided eye. Species that can be caught with a net are called net plankton to distinguish them from the minute varieties that pass through No. 20 silk bolting cloth meshes. The later less than 0.06 mm. in length, include most protozoans, bacteria, and fungal forms collectively called jiannoplankton. Organisms that depend on the surface film for a substratum called neuston and are more important in the quiet waters of ponds than the lakes. Nekton are larger animals that are capable of locomotion independent of water currents. Aquatic birds swim and dive are included in this group. Benthos organisms are attached to or dependent on the bottom for support, there are sessile, creeping and burrowing forms. Periphyton the assemblage of organisms attached to surfaces, either living or non-living, submerged in water above the bottom. Organic detritus, important in terrestrial as well as aquatic ecosystem, is a particulate material originating from disintegrating and decomposing biomass. All these major categories may be sub-divided into smaller units.

BENTHOS

The lake bottom can be divided into *littoral* zone and profundal zone. The littoral zone extends from the water's edge to the limit or rooted aquatic vegetation. It may be sub-divided into the *eulittcal* zone, between high and low water makes at the water's edge, where the beating of waves is most effective, and the sublittoral zone, extends from the lower limit of rooted vegetation. Where such vegetation is absent, the sublittoral zone, may be considered the bottom of the epilimnioh down through the thermocline.

The profundal zone is the entire bottom below the rooted vegetation or commonly the bottom of the hypolimnion. The boundary lines between the zones are variable and change with the depth of thermocline. The open water of the lake above the bottom is known as the limnetic zone.

PLANKTON

Fresh-water plankton includes representative from photosynthetic algae, Bacillariaceae (diatoms), Myxophyceae (blue-green), chlorophyceae (blue-green), and occasional other form such as Wolffia among the higher plants, the non-photosynthetic bacteria and other fungi, and among zooplankton, all classes of protozoa except Sporozoa, Rotatoria, Entomostrabaca (especially Cladocer, Copepoda and Ostracod), some immature, Diptera, the statobiasts and gemmules of bryozoans and sponges, the rare fresh water jellyfish Crospedacusta and occasional aquatic mites, Gastrotrich and others. Fresh water plankton lack many forms common in the plankton of the ocean. On the other hand the rotifers, aquatic insects and water-mites are mostly absent from the sea, and the Cladocera are poorly represented, it is probable that the plankton evolved from benthonic forms occurring near the shore, notably Ostracoda and Rotatoria, are still largely benthonic in behaviour.

NEKTON

The Nekton of lakes consists principally of fish. There is an interesting small shrimp, Mysis Relica, found in the deeper waters of many northern lakes of North America that is often included with the nekton. This species is believed to be relic of a marine fauna that happened to get cut off from the sea in some past geological period, yet was able to survive as waters became fresh.

Numbers and species of fish are more concentrated in the littoral zone of lakes than in the open waters of the limnetic zone.

Limnetic species also invade shallow water for spawning, in deep waters, fish tend to remain close to the bottom, where their food supply is located, unless there is some deficiency of oxygen there. Caged fish, low red to various depths in a eutrophic lake, did not survive long below the thermocline (Smith 1925). Fish may however, make short excursion into the hypolimnion.

In the littoral zone, fish species are segregated according to the composition of the bottom, as are the invertebrates. The species living over rock and gravel bottoms in lakes are mostly different from those inhabiting similar bottoms in streams, but the mud-bottom forms are nearly the same as in ponds.

Amphibians and reptiles do not commonly occur in lakes except around margins supporting attached aquatic vegetation and only pond species occur. Such pond mammals as the muskrat mink, and other are not the typical of lakes as such although they are frequently found in shallow littoral waters. There are a number of bird species, however, that occur most commonly in lakes.

Eutrophication

Eutrophication is a natural process, as oligotrophic lake change to eutrophic ones in the course of centuries, Eutrophication becomes of concern when its rate is greatly accelerate by products of human activities that bring equivalent changes in decades for years. Modern agricultural practices cause increased erosion of uplands and correspondingly siltation of lakes. Excessive organic wastes from cities and farmlands become destructive pollutants ever, in large bodies of water.

Correlated with changes in the physical and chemical characteristics of water there has been a change in the predominant species of phytoplankton; early nineteenth century, *muskellunge, ner'heni pike take trout* were commonly taken in the commercial fisheries but the harvest of these species began to decline rapidly towards the end of the century Cisco, while fish saiigcr blue pike, walleye ling, fresh water cod, and channel catfish were among the principal commercial fish during the first half of the twentieth century but have decreased considerably.

The remedy for situations leading towards accelerated eutrophication, particularly as it rises from the use of detergents, would seen to be either the reduction of the phosphates content in the product, replacement of phosphorus by some other chemical to produce the same clearing effect, return to the use of soap development of special sewage-treatment methods to remove the excess chemicals or a combination of alternatives. There are difficulties with each option. Water treatment technology is improving and phosphorus removals up to 90 per cent or more have been achieved, but conversion of existing sewage-treatment plants to these new techniques will take time and money. Some combination of these different procedures may be required for quick improvement of lake habits.

Reduction of phosphorus from inflows into bodies of water may not always bring a corresponding decrease in excess algae present since at the time the organic growth is controlled by the nitrogen supply rather than the phosphorus. It is even more difficult to remove nitrogen from water effluents than to remove phosphorus. There may be need as with phosphorus for regulation of the use of nitrogen at its source, which is often fertilizer. Certainly cultural eutrophication can be arrested, but this will require ingenuity, time and money.

Ponds, Marshes, Swamps And Bogs

Ponds is a popular terms for lakes of the third order that are small shallow and when mature have rooted vegetation over the bottom. There is no clear distinction between the ponds and lakes of the first and second orders. The littoral zone of eutrophic lakes for instance is pond like in habitat and organisms. Floating and emergent vegetation commonly occurs around the margins of ponds to form extensive tracts of marsh, the pond habitat may originate as shallow basin, as large pool in a stream as the result of the filling in a lake, or from a stream dammed by beaveas, man or landslide. Slow-flowing rivers are essentially elongated ponds and have a similar fauna. Because of the slight water movement in ponds, the surface film becomes an microhabitat for some species, pond water temperature is often uniform at all depths, but during warm sunny weather, ponds protected from the wind may show considerable stratification, not only in temperature, but also oxygen content and other characteristics. Daily and seasonal variations in temperature may be great because of the small volume of water. Ice forms earlier and lasts longer in pond than in lakes, freezing shallow ponds to the bottom in severe winters. Light penetrates to all depths, encouraging growth of vegetation except in high turbidity.

Plant Sere

The plant hydrosere, or pond sere typically contains the following stages and characteristic species: submerged vegetation: Water weed, pondweed, milfoil, hornwort, nalands, buttercup, bladdenvort, eelgrass, and the herb-like algae Chara.

Floating vegetation: Water lily, pondweed, smartweed, duckweed, and water hyacinth, all except the last two are rooted in the mud, often at a depth of 2 to 3 metres and may have rhizomes from which long petioles extend to the leaves floating on the surface. Duckweeds and water hyacinths are unattached floaters, and cover the surface extensively in some localities.

Emergent vegetation (marsh): The dominant species are cattail, reed, wild rice, and saw grass. They invade water of over 1 metre depth, but in shallower waters or in secondary secession they are replaced by the sedge meadow composed of sedge rush and spike rash.

Swamp shrubs: Buttonbush, alder, dogwood, swam rose and sometimes shrubby willow and cottonwood.

Swamp forest: Red and silver maples, elmash, swamp white oak, and pin oak. Succeeding stages depend on the climate region. In arid regions the swamp forest may be poorly developed, and grassland or desert vegetation may come quickly. In the mesic climate of the Eastern stages, on oak-history associes follows the swamp forest, replaced in turn by climax ugar maple-beech or mixed mesophtic forest. The hydrosere in the broad leaved evergreen climax area of south-eastern North American brings in cypress and a number of other unique species.

Animal Sere

Animal characteristic of marshes and ponds constitute a distinct pondmarsh biocies which extends into sluggish or base-levelled streams and the littoral zone of eutropic lakes. Most animal species are restricted to single state or community of the plant sere, but commonly occur in several stages in varying abundance and for activities. Fish, for instance, feed in open water but spawn in shallow water among the emergent vegetation. Submerged, floating and emergent vegetation represent different levels or strain in a single biotic community, and each stratum has about the same degree of distinctiveness as forest community strata. With the invasion of swamp shrubs and the ground well above the water table most of the year, ponds and marsh species largely disappear, replaced by many characteristic new species. The animal community represents the swamp facies of the deciduous forest-edge biocies. The swamp forest is often quite open at its outer margin and forest-edge species remains common. But as this forest develops a closed canopy and dries ground stratum, it is invaded by the swamp facies of the deciduous forest biociation. In the ponds sere there is a succession of animal adaptations from aquatic to amphibious to terrestrial.

Pond Marsh Biocies

Neuston

The supraneuston, organisms which move on top of the surface film in pursuit of most of their life activities, consists of the water strides. Gerridae, Validae, Mesovelidae, the water measures hydrometridae, the whirligig beetles Gyrinidae, the sprintails Coloembola, some spiders and occasional other forms. The gyrinids of several species commonly occur in social groups. They are remarkable for having each eye divided so that the upper portion looks into the air and the lower portion into the water. Several of the forms listed have long legs that distribute the weight of the body over a large area of surface of film. The portions of the legs or body that contact the surface film are water repellant.

The under surface of the water film supports an infraneuston of Hydra, planarians, obstracods, dadocerans, snail's and insect's eggs, larvae, and Pupae (mosquitoes) certain kinds of midge flies, and so forth. For all except the insects, however, the use of surface film in this manner is usually transitory. Some cladocerans such as Bosmina and Baphnia occasionally breakthrough the surface film from below, fall over into their sides and cannot return.

Plankton

The species of plankton found in ponds differ somewhat from those in lakes but the transition from lake species to pond species is gradual and progressive one. Protozoa and Rotatoria are usually more abundant in ponds than in lakes.

Although plankton distribution does not vary with depth as much in ponds as in lakes, seasonal fluctuation are similar in nature.

Benthos

Subaquatic animals dwell not on the bottom but also on the stems and leaves of submerged plants. Aquatic plant species have diversified animal periphyton of insects, amphipods, mites and snails, using them for food, shelter, or reproductive sites denied them in mud bottom below. Other kinds of insect larvae and oligochaetes are more abundant in the mud than on the plants. The under surface of lily pads often contains many small organisms, including Protozoa, Hydra, Flatworms, Rotifers and snails. The biomass of animals varies directly with the biomass of vegetations, and the quantity of invertebrates is especially great on those plants possessing finely dissected leaves. Very few species found in lake bottoms are found in ponds, but the pond-marsh biocies contains many species not found in lakes.

The number of species and individuals found in the bottom fauna increases with the age of the pond from the time the pond is formed till the attached vegetation becomes excessive. Coincident with the development of the bottom fauna is an increase in variety and abundance of plankton and fish.

Terrestrial Invertebrate

The terrestrial insects found in marsh vegetation are in the main adult mosquitoes, midges, dragonflies, demselflies, mayflies and alderflies, whose immature stage like submerged. On bare ground around ponds may be found tode bugs, shore bugs, spring tails, tigen beetles, and sometimes ground beetles and pigmy locusts. Spiders become numerous throughout the vegetation, and the snail Succimca appears. In addition to these marsh and ponds species invertebrates belonging to the forest-edge biocies may occasionally be found.

Fish

Fish are often abundant, included among the species that occur are several bottomfeeders-suckers, bullheads, buffalo and carp. Feeding on the submerged vegetation and stirring up the bottom, they may control the habitat and the composition of both animals and plant species present in the community. This condition, however, does not last indefinitely. Vegetation encroaches on the margins of ponds, and the fish are gradually eliminated because of the disappearance of suitable breeding sites. The mudminnow, bowfin, and bulhead are usually the last to disappear before the pond becomes a dry marsh.

Amphibians and Reptiles

Salmanders and frogs are basically aquatic animals, although they show varying degrees of adaptation to terrestrial life. Siren and Necturus have permanent, external gills and spend all their live in the water. Most other forms lay their eggs and pass through their early development in water, but the adults are airbreathing and wander over the land. Since their skin must be kept moist, they are confined to the vicinity of water, to humid climates, or to damp humus.

Reptiles are Terrestrial

Desert reptiles never go to water, Painting, geographic, and snapping turtles bask in the sun on the shore or on protruding logs, but quickly plunge into the water to escape danger, to cool off, or to feed. The alligator, musk, and softshelled turtles spend nearly all their time in water. The softshelled turtle is able to utilize dissolved oxygen in the water and hence has evolved special readaptation to water. Like other turtles, however, they lay their eggs on land, placing them in holes excavated in sand, loose soil, muck, or decaying stamps or logs. Water snakes give birth to living young ones that enter the water immediately. Water snakes feed on insects, small fish and amphibians, crayfish, or whatever other animal food they can find.

Birds

Birds population are high, and the pondmarsh, swamp shrub or forestedge, and forest communities are especially clearly defined. Both the number of species and the equitability index increases with succession. There is an abundance of nest-sites and food, but the aquatic and terrestrial species exploit different niches to avoid competition as much as possible. The aquatic species feed in all stages of the plant sere, beginning with the open water, but nest for the most part in the emergent vegetation.

Food Chains In Ponds

Many animals in ponds depend for food on floating phytoplankton, bacteria and bottom detritus. Filamentus algae are more abundant in ponds than lake and are a source of food to various immature insects are from tadpoles. Ponds, unlike lakes have additional producers of organic matter in the rooted pondweeds. Pondweeds are consumed by insects, ducks, and herbivorous mammals. The periphyton of bacteria diatoms, green and microscopic animals is important as food to many small crustaceans, immature insects, oligochaete, worms, and snail. Creeping predators among the pondweeds and alage are leeches, dragonfly and damselfly and naiads, and water mites small swimming predators are the dytiscid beetles, most of the hemipterans, and the swimming leeches Erpobdella and Macrobdella. At the top of the food chains, feeding on all these small animals, and often on plants as well, are the fish and other vertebrate groups.

Mammals

One of the most characteristic mammals of the marsh is the muskrat. A well developed marsh may contain one of their haycockshaped on each acre (2.5 per hectare), with perhaps five animals per lodge during the autumn. The diet of the muskrat is largely the leaves and roots of marsh vegetation although they also feed to some extent on crayfish, clams, snails, and sluggish fish. Overpopulation of two or three lodges per acre (5.0 to 75 per hectare) may lead to 'eat-outs' or local destruction of the marsh vegetation.

The mink is probably the most common mammalian predator of the marsh, it is an enemy of the muskrat. Foxes, recalls, and coyotes may invade the marsh when the water level is low. The otter preys on fish and fish on the marsh, it has now been exterminated from most of its former range.

The breaver makes its own marsh habitat by damming small streams, flooding the surrounding lowland. Here it builds its large lodge and feeds on the bark and twigs of adjacent aspen, willow and cottonwood, and on the roots of aquatic plants. When supply of aspen and other food is exhausted, the colony disappears, the dam decays, the water level subsides and marsh vegetation invades, after some years the pond is converted into a beaver meadow.

Bogs

Bogs or moors typically develop in the hydrosers of cold northern regions, while marshes and swamps, which are markedly different from bogs, are characteristically southern in their location. Bogs commonly develop into a coniferous forest climax, swamps succeed to deciduous forest at other southern climax types. Several thousand years ago, when glacial climate grippe the northern states, extensive bogs developed and have persisted as relic communities in spite of the warming of the climate. These bogs are slowly being replaced by pondmarsh species at equivalent serai stage by cliseral succession.

Bogs occurring in the Great lakes region are ordinarily small in area and have little or on drainage. There may be oxygen present in the open water of the large bogs, but it is characteristically in very low concentration, at the most in small bogs or in the marginal zones. Bog water has a distinct brown colour a low nitrogen content, a low temperature beneath the surface a low pH at least in the marginal vegetated zones, and a low dissolved salt content.

A false bottom is characteristic of bogs. It consists of finely divided plant material of a light brown colour, head suspended in the water at varying depths below the surface. This false bottom may be extended downward several meters before a true solid bottom is reached. The material disperses on slight disturbance and may render all the open water turbide. Ordinarily, the surface waters are quiet and clear. Dead vegetation does not completely decompose, as accumulates, it becomes compressed to form peat.

Plant Bog Sere

In the early stage of development of the bog, organic detuitus may accumulate mostly in the deepest portions. In some bogs the first plant stage may be composed of floating vegetation, but floating vegetation is often absent and the first stage is a sedge-mat composed of sedges cotton-grass, and buckbean. The rhizomes of the sedges grow out into the water and become so interlaced that they form a floating mat.

The next plant stage is dominated by low shrubs, which encroach on the floating mat. The leather leaf, bog rosemary, laurels, labrador tea, sweet gale and cranberries are important species.

A high shrub stage commonly follows the low shrubs at such times as the mat becomes thicker or grounded. Common shrub species are holly, chokeberry, alders and dwarf birch.

The first tree of the bog forest to invade the shrubs in commonly the tamarack, but this species is now less common than formerly because of fire, logging, and the deredation of the sawfly larvae Lygaeonewaus ericisoni. Black spruce may either invade the shrubs directly or follow the tamaracke. Later, the northern white cedar may become dominate and persist for a very long time, but the ultimate fate of the bog, upon addition of upland soil or the water table which in to cover with the climax forest of the region.

Animal Life

In bogs that have a large body of open water, or on inflow of water entraining oxygen, and in which the pH is not extreme, invertebrate life comparable to that found in ponds and marsher occurs. True bogs, however, have little oxygen and a low pH, and many pond species do not appear. Mollusks are characteristically absent, sphareriids may persist but their shells become very thin. Bottom organisms in general are poorly presented because of the tenuous physical nature of the substratum. Desmids predominate among the phytoplankton, although dinoflagellates, Chlorophyceae, and Myxophyceae are common. Rotifer and a variety of Protozoa are the principal zooplanktons.

Amphibians and reptiles are not characteristic of bogs, although the leopard frog is sometimes numerous on the sedge mat of bogs.

In general, the productivity and economic value of bogs is very low compared with ponds and marshes. Liming experiments, calculated to improve productivity, are being made. Calcium combines with the humid colloids, which then flocculated and fall to the bottom. This clears the waters, light penetrates deeper, pH is raised and greater algal, zooplankton, and fish growth is induced. Peat is a special product of bogs. It is of importance in northern Europe. It is cut out in blocks, dried, and used as fuel and for mixing in the soil of gardens and lawns.

Rock, Sand And Clay

The origin of life was undoubtedly in the sea. Physiological adjustments were necessary before organisms could occupy freshwater. Although some organisms may have become air-breathers and invaded land habitats directly from the sea, most evolution of terrestrial forms has doubtless come from fresh water. Relatively few major groups of animals have been successful in this invasion of land, the most notable being oligochaete worms, gastropod, mollusks, many, arthropods, especially the insects and spiders, reptiles and their evolutionary offspring, the birds and mammals.

Adjustments To The Terrestrial Habitat

Living on land presents many problems. The present concern is to analyse the ways in which animals have met these problems and to trace the succession of communities in the extreme terrestrial habitats of rock, sand, and clay. A variety of techniques have been developed for measuring environmental factors.

Gravity

In water, organisms counteract gravity by means of various floating and swimming mechanisms. Fluid buoyancy permits waterdwellers to attain huge size, consider the whale. A land animal, on the other hand, must support its entire weight. Some terrestrial animals gain a modicum of support by burrowing into the soil. Others drag their bodies over the grounds surface. But the supportive advantage they thus gain is costly in other directions, for they are slow moving and relatively helpless before predators. The animals best adapted to terrestrial life have evolved, appendages in the forms of legs or wings that not only raise the body above the ground but are also the means of more or less rapid locomotion an adroit movements over the surface of through the air. Terrestrial adaptation has involved the development of a tough body covering to hold fluids and internal organs in place, skeletal framework to give permanent shape to the body and, as system of levers, to furnish means of locomotion, and powerful muscles to lift and move the heavy body.

Moisture

In sharp contrast to aquatic forms, terrestrial animals are not constantly enveloped with a continuous watery medium, with the limited exceptions of protozoans, nematodes, and other small organisms living in moist soil. Land animals, lacking constant with the water medium, are faced with the problems of obtaining water and preventing excessive water losses from body.

Water becomes available to animals in varying amounts in the forms of rain, snow, hail, frost, and fog. Whatever the form it arrives in, the significant things are the amount of free liquid water added to the substratum, accessible to plants and animals, and the humidity of the air. Considerable amounts of moisture are lost to organisms as runoff water flowing into streams, by evaporation back into the air, and as water bound in snow and ice.

Water is cast from the body through the skin and lungs as insensible moisture perspiration. Rapid and largely uncontrolled loss of moisture through the skin of amphibians, snails, anelids, and insect larvae is limiting factor confining these animals to mist habitats or to activity only in times of high humidity. Adult insects and other arthropods, reptiles, birds and mammals have evolved body surfaces of chitin and waxes, scales, or comification of the surface layers of the skin that largely prevent uncontrolled loss of moisture. Moisture loss through the respiratory surfaces in these forms remains considerable. Water is also lost with the faces, although in some species much water is reabsorbed by the large intestine before the faces are ejected. The amount of water removed from the body by excretory organs, particularly the kidneys. The kidneys are critical to maintenance of proper concentration of salts in the blood and body fluids. Organisms are very sensitive to disturbances in body water balances and this factor is significant in determining the types of niche which a species comes to occupy.

In order that animals could exist in terrestrial habitats, they had to acquire the ability to carry on reproductive activities in absence of water. The chief reproductive adaptations involve the following internal ferruination, a shell covering the egg to conserve moisture and salts, food provision to the embryo and young, by yolk in the egg cell, placenta in the uterus of the mother, of direct feeding, by adults, reduction in number of young with more efficient parental care, reduction or elimination of free-swimming larval stages, and greater segregation of species into different niches to avoid interspecific disturbances.

Temperature

Aquatic animals are not ordinarily subject to temperature below freezing point and are in a relatively stable temperature environment, but terrestrial species are exposed to highly variable temperature that may reach extremes of about 68°C. No single species is required to withstand such as wide range of temperatures, however, optimum and tolerance limits vary from one species to another.

No aquatic species has evolved control over its body temperature. Aquatic warm-blooded mammals and birds are derived from terrestrial forms. An ability to maintain a constant body temperature has survival value in terrestrial habitats, however, and consequently physiological mechanisms for homoiothermism developed independently in birds and mammals. All other land organisms are poikilothermal, that is they have no physiological mechanism for maintaining a constant body temperature. Some poikilotherms such as bees, have developed special behaviour patterns that enable them to maintain fairly constant conditions in the hive by cooperative efforts, some moths and other insects with hair-like body coverings maintain a degree of constancy in body temperature by shivering their wings, some lizards, snakes, and turtles are able to exert some control over their body temperatures by moving into and out of sunlit areas.

Rates of activity, food consumption, metabolism, growth and other physiological functions increase, to a certain limit with rise of body temperature. Homoiotherms maintain a continuous high rates of functioning because of their constant high body temperature, but the rate at which poikilotherms function varies with the temperature of their habitats.

Oxygen

Dry air at 760 mm Hg pressure contains approximately 21 per cent oxygen, 0.03 per cent carbon dioxide, 78 per cent nitrogen and traces of other gases. Oxygen is thus much more abundant, constant and available at all times in air that it is in water. Oxygen availability seldom becomes a critical factor for land animals, with occasional exception of forms that live in the soil or invaded high altitudes.

Although terrestrial organisms have evolved simple most chambers, branched trachea systems, or complicated lungs to replace the gills found in many aquatic forms, the fundamental requirement of moist membranes for the exchange of oxygen and carbon dioxide between body fluids, tissues, and surrounding medium remains the same. The skin still serves this purpose in some terrestrial forms-annelids and some amphibians-but in most forms the moist membrane are within the body. Internal placement decreases the loss of water through evaporation. The evolution of an ability to take oxygen directly out of the air apparently preceded the actual invasion of land,·and may have been induced invasion in the pond and marsh habit when oxygen dissolved in the water becomes reduced or absent during summer stagnant periods. The evolution of internal air-breathing organs was probably concurrent with the evolution of mechanisms to prevent excessive water loss from the exposed surfaces of the body.

Rock Sere

Stages in the plant sere on bare rock include lichens, mosses, herbs and grasses, shrubs, and forest. The special composition of each stage varies with the chemical nature of the rock, the prevailing climate, and the locality.

In the first stage, various kinds of lichens compete for a foothold, but crustose types usually precede foliaceous types. Mosses such as fruticose lichens as Cladonia foliaceous lichens, or may initiate the sere telescoping the earlier lichen stages.

Lichens and mosses soak up moisture in wet weather. They derive mineral nutrients from the underlying rock. Carbon dioxide secreted from the rhizoids from a weak acid with water and dissolve the binding material of the small rock particles. Rhizoids may penetrate rock for several millimetres. These plants trap wind blown dust and obtain nitrogen from organic compounds in it. When the plants die, they become an addition to the accumulation of organic matter. Herbs, grasses, ferns and later stages invade to continue the crumbling of the rock and buildup of the soil.

Animal Life

Animal life in the pioneer stages on rock is scanty. Ants and spiders roam over the bare rock, and insects of various sorts may stop there, temporarily. Mosses offer a somewhat more substantial micro-habitat to those animals that can tolerate great extremes of flooding, dessication heat and cold can survive such forms in the rhizopod protozoans, nematodes rotifers, copepods, small insects and mites. They often have special means of attachment to keep them from being blown away by the wind, such as strong claws or cement glands on the feet, long bristle-like threads to entangle among the mosses filaments stickers of spines covering the eggs. Since wet periods are often too short to permit complete development, all stages must be tolerant of dessication, at which time activities and growth are largely suspended.

Animal life in general and land snails in particular are usually more abundant in vegetation (grassland and forests) established on calcareous soils derived from limestone than in the vegetation established on soils derived from sandstones, granite or volcanic rock.

Clay Sere

Erosion or calculated removal of overlying material in strip mining for coal may leave bare areas of clay exposed and impound small bodies, water, often with low values. These water bodies undergo a modified pond succession. In clay above pH 4.5 annual plants, of which smartweed is particularly important, appear within a few week to two years, the higher the clay pH, the quicker the appearance of vegetation. Within 2 to 5 years thereafter sweet clover invades and develops nearly complete dominance over large areas. Sweet clover is a biennial, and an exotic species unimportant in the sere in some parts of the country. Prior to its introduction, this stage in the sere on bare clay may have consisted of the perennial grasses still found in small scattered patches. The first trees begin to invade early in the sere, out they are scattered and slow of growth, and do not attain dominance for 25 to 30 years. Willows occur in wet spots. Herb species of the first two plant stages disappear, for the most part, in the shrub stage. The herb stratum now consists largely of wood nettle. Advance forest stages of oake, hickories, basswood, and sugar ample will likely invade in the future, as they occur now in adjacent areas.

Animal Life

The number of invertebrate species tends to increase as the sere advances, although not always regularly. In a study of a formerly stripmined area, 18 species were found to be important in the annual stage, 41 species in sweet clover, 40 species in shrubs, 32 species in the early forest stage, and 67 species in the upland climax. More species would be found in advanced stages because of the greater variety of niches available. Thus, in the initial bare area there is only the ground stratum, in the annuals and sweet tree strata. Since the early forest is on a floodplain, the ground is frequently swept by floods, and the shrub stratum is poorly represented. The climax forest has all strata, richly developed, and possesses the greatest number of animal species. There is an increase in the abundance of individuals per m^2 with the progression of the sere.

Starting with bare ground and continuing through early and the shrub stages to bottom land forest, there is a progressive increase in bird species diversity. The first small mammal to invade the areas is the Praise deer, mouse followed by cottontail rabbit and fox-squirrel.

GRASSLAND, FOREST AND FOREST EDGES

The succession of animal communities in humid climates passes through three terrestrial stages before attaining climax: grass shrubs and scattered trees (forest-edge), and forest. In arid climates the climax may be reached at first or second stage.

VEGETATION

Grassland vegetation differs from forests in that the above-ground vegetation is completely renewed each year. Grasses may be divided into three categories on the basis on height tall grasses (1.5-3 m tall), such as big bluestem and slough grass; mid-grasses (0.5-1.5 m tall), such as little bluestem and needle grass; and short grasses (less than 0.5 m tall), such as buffalo grass. The taller grasses grow in wet habitats, the short grasses in arid habits. Most native grasses are bunch grasses in that they grow in clumps with the areas between the clump either bare ground or occupied by the species. Broad-leaved herbs occurring between the dominant grasses are called forbs.

Forests are composed of trees growing sufficiently close together to dominate the entire area of ground surface. In cold climates most forest are needle leaved evergreen, in seasonally warm moist climates, they are mostly broadleaved deciduous, and in continuously warm, moist climates, they are broadleaved evergreen. In spite of these secondary differences in life-form, the structure and internal dynamics of all forest communities are quite similar.

Between forests and open country, scattered trees are interspersed with shrubs and grasses. This transition area is usually narrow around the margins of a stabilized forests, but where succession is occurring, large areas of mixed grass, shrubs, and trees may provide in dose proximity a variety of habitats which animals take advantage of in a variety of unique ways. Likewise, in agricultural areas, hedge and fence rows, or narrow strips of trees and shrubs along streams, are really edges without the adjacent forest and have the same characteristic animal life.

The grassland community typically contains subterranean, surface and herb strata of vegetation. The forest in addition has shrub and tree strata. Correlation with the differences in microclimates both plant and animal species show segregation of their activities and presence to one or more of these strata.

The herb stratum in mature grassland is much better enveloped in biomass than in the forest, but the woody vegetation of the forest is so great that the total biomass of the community is considerably greater. In one study, it was 29 times greater. Good comparative data on the biomass of all animals are not available, but class studies in central illinois during early autumn indicate no great differences, in the macroscopic invertebrates. Serai grassland communities may, however, in contrast to climax ones, be relatively impoverished.

Grassland Community

Snails, earthworms, and myriapods are not numerous in most grasslands because of the dry habitat. Insects, however, are abundant. Spiders make up about 7 per cent of the total arthropod population in grassland.

Birds are not numerous in grassland. In north-western Towa, grasshopper sparrows western meadowlarks, bodolinks, rigenecked peasants, marsh hawks, and owls averaged less than one pair per hectare (25 acres), Prairies chicken and sharp-tailed grouse.

Several species of snakes are generally distributed over the prairies. To the east, the blue racer, messasuga, bullsnake, and garter snakes are frequently found. The lizards like cremidoplwurs, sexieneatus its commonly occurs in grassy areas at forest edges.

Forest Edge Community

Grassland animals are usually restrained from penetrating forests in the same way that true forest animals are restrained from penetrating grassland, although the home ranges of these species may overlap at the forest margin and in shrubby areas. Since shrubs are especially areas. Since shrubs are especially numerous at the forest-edge and animals have an opportunity to make use of these as well as both grassland and forest, the forest-edge biocies is well-developed for some groups of animals.

Forest Community

Since the censusing of each group of animals furnishes special problems, there have been no studies of total animal populations in single forest communities. The total measofauna would number in the tens or millions and the microfauna in numbers so large as to be scarcely conceivable. The biomass of bacteria would probably be somewhat less than that of fungi but exceeding that of algae, protozoans, and nematodes combined. In general the number of individuals representing a species varies inversely with the body size characteristic of the species. There is, however, considerable variation in population levels both geographically and temporarily.

Stratification

Animal life in the community is separated into different niches in relation to strata. Food, shelter, and microclimatic differences are the chief limiting factors. Because microclimate in each stratum varies from hour to hour, day to day and season to season, classification of a species by stratum must be in terms of the stratum it is observed to frequent for the major portion of a relatively long period of time. The inhabitants of the five strata divide into two major groupings, or societies. The soil invertebrates and some mammals move freely back and forth between subterranean and surface strata, and may be considered a ground society, distinct from a foliage society which occupied the combined herb, shrub, and trees strata. The two societies are not mutually exclusive for ground animals such as millipeds and snails climb up onto herbs and tree trunks during humid weather and foliage animals often rest on the ground, and search for food, hibernate, and lay eggs, there.

Population density of invertebrates ranks highest in the surface stratum, followed by subterranean and herb strata, while shrub and tree strata rank lowest. The largest mammal populations occur in the subterranean and ground strata.

Seasonal Changes

The forest community changes drastically with the season such that six aspects may be recognised. The total population of the soil macrofauna in the temperate deciduous forest is the highest during the winter because of the migration of many foliage insects into this stratum to hibernate. Forest species hibernate in densities that vary randomly throughout the forest, except where are differences in topography or substratum. Forest tracts adjacent to grass or farmland, however, receive an influx of non-forest species that hibernate principally on the forest edge generally, and along the southedge in particular, where exposure to solar radiation and protection from cold, northerly winds produces warmer temperature.

During the general concept, insects and other invertebrates come out of hibernation and the adults of the varieties of herbs, shrubs and tree foliage grow and return to the stratum. The population of ground animals remains relatively at higher stage throughout the year and depends on geobionts for reproduction. Various group of these geobionts and geophils reach peak in numbers of adults at different times during the year.

Bird population in temperate zones reach peak in its populations with the passage of transient during the season of migration. In tropics, birds are most abundant during the winter period of the north temperate zone, since the fauna then contains many migrant species from the north.

Pic. 3.1: A deceptive beauty—the Fountain Hills in Phoenix, Arizona

Pic. 3.2:"Natural World : The Tiger's Fortrees" on Animal Planet, Wednesday, 20.00.

4

THE WORLD ENVIRONMENT SCENARIO

Over the course often thousand years humans have successfully learned to exploit ecological system for sustenance. While ecological systems are supple, they can snap viciously when bent too far. The land's ability to serve human ends can be markedly and sometimes permanently snapped. Humans out of ignorance, short-sightedness, greed or desperation have polluted air and water, undermined the productivity of the land through accelerated soil erosion, creeping deserts, increased flooding and declined soil fertility. They, thus destroy the basis of their own livelihood and they violate the limits of natural systems. The principal victims of these trends are the world's poor, who, in their quest for food and fuel, are often forced by circumstances beyond their control to serve as the agents of their own undoing.

Stockholm Conference

The global concern for environmental protection found expression in the UN conference on the Human Environment, held at Stockholm in June 1972. The Conference adopted an Action Plan relating to subject areas of human health, territorial ecosystems; environment and development; ocean's, energy; and natural disasters, it proposed institutional arrangements for carrying out the Action Plan within the U.N. system. The recommendations of the Action Plan were accepted by the General Assembly and institutional arrangements made under United Nations Environment Programme.

The United Nations Conference on the Human Environment

Having met at Stockholm from 5 to 16 June, 1972.

Having considered the need for a common outlook and for common principles to inspire and guide the peoples of the world in the preservational and enhancement of the human environment.

I. Proclaims That

1. Man is both creature and moulder of his environment, which gives him physical sustenance and affords him the opportunity for intellectual, moral, social and spiritual growth. In the long and tortuous evolution of the human race on this planet a stage has been reached when, through the rapid acceleration of science and technology, man has acquired the power to transform his environment in countless ways on an unprecedented scale. Both aspects of man's environment, the natural and the man-made, are essential to his well-being and to the enjoyment of basic human rights even the right to life itself.
2. The protection and improvement of the human environment is a major issue which effect the well-being of peoples and economic development throughout the world; it is the urgent desire of the people of the whole world and the duty of all Governments.
3. Man has constantly to sum up experience can go on discovering, inventing, creating and advancing. In our time, man's capacity to transform his surroundings, if used wisely, can bring to all people the benefits of development and the opportunity to enhance the equality of the life. Wrongly or

needlessly applied, the same power can do incalculable harm to human beings and the human environment. We see around us growing evidence of man-made harm in many regions of the earth: dangerous levels of pollution in water, air, earth and living beings; major and undesirable disturbances to the ecological balance of the biosphere destruction and depletion of irreplaceable resources; and gross deficiencies harmful to the physical, mental and social health of man, is the man-made environment, particularly in the living and working environment.

4. In the developing countries most of the environmental problems are caused by underdevelopment. Millions continue to live far below the minimum levels required for a decent human existence, deprived of adequate food and clothing, shelter and education, health and sanitation. Therefore, the developing countries must direct their efforts to development, bearing in mind their priorities and the need to safeguard and improve the environment. For the same purpose, the industrialized countries should make efforts to reduce the gap between themselves and the developing countries. In the industrialized countries, environmental problems are generally related to industrialization and technological development.

5. The natural growth of population continuously presents problems on the preservation of the environment, and adequate policies and measures should be adopted, as appropriate, to face these problems. Of all things in the world, people are the most precious. It is the people that propel social progress, create social wealth, develop science and technology and through their hard work, continuously transform the human environment. Along with social progress and the advance of production, science and technology, the capability

of man to improve the environment increases with each passing day.

6. A point has been reached in history when we must shape our actions throughout the world with a more predent care for their environmental consequences. Through ignorance or indifference we can do massive and irreversible harm to the earthly environment on which our life and well-being depend. Conversely, through fuller knowledge and wise action, we can achieve of ourselves and our posterity a better life in an environment more in keeping with human needs and hopes. There are broad vistas for the enhancement of environmental quality and the creation of a good life. What is needed is an enthusiastic but calm state of mind and intense but orderly work. For the purpose of attaining freedom in the world of nature, man must use knowledge to build, in collaboration with nature, a better environment. To defend and improve the human environment for present and future generations has become an imperative goal of mankind-a goal to be pursued together with, and in harmony with, the established and fundamental goals of peace and of world-wide economic and social development.

7. To achieve this environmental goal will demand the acceptance of responsibility by citizens and communities and by enterprises and institutions at every level, all sharing equitably in common efforts. Individuals in all walk of life as well as organizations in many fields, by their values and the sum of their actions, will shape the world environment of the future. Local and national governments will bear the greatest burden for large-scale environmental policy and action within their jurisdictions. International cooperations also needed in order to raise resource to support the developing countries in carrying out

their responsibilities in this field. A growing class of environmental problem, because they are regional or global in extent or because they affect the common international realm, will require extensive co-operation among nations and action by international organizations in the common interest. The Conference calls upon Governments and peoples to exert common efforts for the preservation and improvement of the human environment, for the benefit of all the people and for their posterity.

II. Principles

State the common conviction that :

Principle 1

Man has the fundamental right to freedom, equality and adequate conditions of life. In an environment of a quality that permits life of dignity and well-being, and bears a solemn responsibility to project and improve the environment for present and future generations. In this respect, policies promoting or perpetuating apartheid, racial segregation, discrimination, colonial and other forms of oppression and foreign domination stand condemned and must be eliminated.

Principle 2

The natural resources of the earth including the air, water, land, flora and fauna and especially representative samples of natural ecosystem must be safeguard for the benefit of present and future generations through careful planning or management, as appropriate.

Principle 3

The capacity of the earth to produce vital renewable resources must be maintained and, wherever practicable, restored or improved.

Principle 4

Man has a special responsibility to safeguard and wisely manage the heritage of wildlife and its habitats which are now gravely imperilled by a combination of adverse factors. Nature conservation including wildlife must therefore receive importance in planning for economic development.

Principle 5

The non-renewable resources of the earth must be employed in such a way as to guard against the danger of their future exhausting and to ensure that benefits from such employment are shared by all mankind.

Principle 6

The discharge to toxic substance or of other substances and the release of heat, in such quantities or concentrations as to exceed the capacity of the environment to render them harmless, must be halted in order to ensure that serious or irreversible damage is not inflicted upon eco-systems. The just struggle of the peoples of all countries against pollution should be supported.

Principle 7

States shall take all possible steps to prevent pollution of the seas by substances that are liable to create hazard to human health to harm living resources and marine life, to damage amenities or to interfere with other legitimate uses of the sea.

Principle 8

Economic and social development is essential for ensuring favourable living and working environment for man and for creating conditions on earth that are necessary for the improvement of the quality of life.

Principle 9

Environment deficiencies generated by the conditions of underdevelopment and natural disasters poses grave problems and can best be remedied by accelerated development through the transfer of substantial quantities of financial and technological assistance as a supplement to the domestic effort of the developing countries and such timely assistance as may be required.

Principle 10

For the developing countries, stability of prices and adequate earnings for primary commodities and raw material are essential to environmental management since economic factors as well as ecological processes must be taken into account.

Principle 11

The environment policies of all States should enhance and not adversely affect the present or future of developing countries, nor should they hamper the attainment of better living conditions for all, and appropriate steps should be taken by states and international organizations with a view to reaching agreement on meeting the possible national and international economic consequences resulting from the application of environmental measures.

Principle 12

Resources should be made available to preserve and improve the environment, taking into account the circumstances and particular requirements of developing countries and any costs which may emanate from their incorporating environmental safeguards into their development planning and the need for making available to them, upon their request, additional international technical and financial assistance for this purpose.

Principle 13

In order to achieve a more rational management of resources and thus to improve the environment, states should adopt an integrated and co-ordinated approach to their development planning so as to ensure that development is compatible with the need to protect and improve the human environment for the benefit of their population.

Principle 14

Rational planning constitutes an essential too for reconciling any conflict between the needs of development and the need to protect and improve the environments.

Principle 15

Planning must be applied to human settlements and urbanization with a view of avoiding adverse in this respect, projects which are designed for colonialist and racist domination must be abandoned.

Principle 16

Demographic policies, which are without prejudice to basis human rights and which are deemed appropriate by Governments concerned, should be applied in those regions where the rate of population growth or excessive population concentrations are likely to have adverse effects on the environment or development, or where low population density may prevent improvement of the human environment and impede development.

Principle 17

Appropriate national institutions must be entrusted with the task of planning, managing or controlling the environmental quality.

Principle 18

Science and technology, as part of their contribution to economic and social development, must be applied to the identification, avoidance and control of environmental risks and the solution of environmental problems and for the common good mankind.

Principle 19

Education in environment matter, for the younger generation as well as adults, giving due consideration to the underprivileged, is essential in order to broaden the basis for all enlightened opinion and responsible conduct by individuals, enterprises and communities in protection and improving environment in its full human dimension. It is also essential that mass media of communications avoid contributing to the deterioration of the environment, but, on the contrary, disseminate information of an educational nature, on the need to protect and improve the environment in order to enable man to develop in every respect.

Principle 20

Scientific research and development in the context of environment problems, both national and multinational, must be promoted all countries, especially the developing countries. In this connection, the free flow of up-to date scientific information and transfer of experience must be supported and assisted, to facilitate the solution of environmental problems; environmental technologies should be made available to developing countries on terms which would encourage their wide dissemination without constituting an economic burden on the developing countries.

Principle 21

State have, in accordance with the Charter of the United Nations and the principles of International law, the sovereign right to exploit their own resources pursuant to their own environmental policy, and the responsibility to ensure that activities within their jurisdiction or control do not cause damage to the environment of other state or of areas beyond the limits of national jurisdiction.

Principle 22

States shall co-operate to develop further the international law regarding liability and compensation for the victims of pollution and other environmental damage caused by activities within the jurisdiction or control of such States or areas beyond their jurisdiction.

Principle 23

Without prejudice to such criteria as may be agreed upon by the international community; or to standards which will have to be determined nationally, it will be essential in all cases to consider the systems of values prevailing in each country, and the extent of the applicability of standards which are valid for the most advanced countries but which may be inappropriate and of unwarranted social cost for the developing countries.

Principle 24

International matters concerning the protection and improvement of the environment should be handled in a cooperative spirit by all countries, big or small, on an equal footing. Co-operations through multilateral or bilateral arrangements or other appropriate means is essential to effectively control. Prevent, reduce and eliminate adverse environmental effects resulting from activities conducted in all spheres, in such a way that due account is taken of the sovereignty and interests of all States.

Principle 25

States shall ensure that international organizations play a coordinated, efficient and dynamic role for the protection and improvement of the environment.

Principle 26

Man and his environment must be spared of the effects nuclear weapons and all other means of mass destruction. States must strive to teach prompt agreement, in the relevant international organs, on the elimination and complete destruction of such weapons.

An assessment after ten years of the Stockholm Conference, suggested that man's perception of the natural world had changed. Notable advances had taken place in the scientific means of examining the complex and interrelated systems of the biosphere, and space research had provided means for large scale monitoring of the earth, its resources and its processes.

Increased emphasis had been placed during the decade on environmental education and public awareness with non-governmental organisation playing an important role, consequently, environmental issues have come to be better perceived. It is not appreciated that all environmental systems are closely interconnected, and are subject to natural and man-made changes which in some cases are irreversible.

There has been a massive expansion in the establishment of national environmental bodies and in legislation dealing with environmental issue. Good amount of work has been done also through the United Nations system. With the establishment of the United Nations Environment Programme

(UNEP) in 1982 and UN Centre for Human Settlement (UNCHS) later, the environmental activities have been receiving a sharper focus. Various specialised agencies on the UN system such UNDP, UNIDO, FAO, WHO and WMO (World Meteorological Organisation) have also given added emphasis in their programme on issues of environmental concern.

However, the overall state of the natural environment has not improved; indeed, deterioration is proceeding at accelerated rates in a number of its components.

UNEP Session Of Special Character

In May 1982, UNEP convened in Nairobi a Session of a Special Character (SSC) of the Governing Council of UNEP to commemorate the tenth anniversary of the Stockholm Conference of the Human Environment. The Session reviewed major achievements of the Stockholm Action Plan and recommended action in the field of the environment to be taken by UNEP over the next 10 years.

The Conference adopted a "Nairobi Declaration" and decision "Environment in 1982-Retrospect and prospect". The Conference felt that the principles of declaration of the Stockholm Conference were as valid today as they were in 1982 and together with the decisions adopted in the Nairobi Conference provided basic guidelines for effective and sustained environmental progress.

The Conference identified the basic orientation of the United Nations Environment Programme for 1982-1992. It charged UNEP to focus its attention on assessment of environmental problems of world wide concern; promotion of appropriate programmes for resource and environmental managemental management as an integral part of development with particular attention to the needs of developing countries, and promotion of environmental law and guidelines.

The Conference suggested that in pursuing these objectives UNEP should be guided by major environmental trends, potential problems and priorities of action (Kindly refer Annexure) and should concentrate in particular on:

(i) Promotion of land and water management, including control of desertification and deforestation;

(ii) Promotion of natural resources;

(iii) Promotion of the International Drinking Water and Sanitation Decade;

(iv) Promotion of new and renewable sources of energy;

(v) Promotion of regional seas programmes;

(vi) Prevention of environmental disturbances from air pollution;

(vii) Prevention of chemical safety and control of hazardous substances.

The Conference asked all governments, inter-governments organisations and non-governmental organisations to ensure that the priorities for action was accorded high priority within respectic programmes.

World-wide Environmental Issue

The salient features of the world environmental issues are described below:

(a) Climatic changes/Green house effect

Many studies show that the atmospheric pollution is on a global scale. Today, however, it is not only the air over our cities that is polluted, the entire atmosphere of our planet is now afflicted to some degree. Meteorologists talk about a nebulous veil of air pollution encircling the entire earth. Smog has been observed over oceans, over the North pole, and in their unlikely places. There are range of environmental pollution issues originating largely in developed countries which are also of global significance. Carbon dioxide (CO_2) build-up leading to climate change is one of those key issues. Recent research indicates that increasing concentrations of CO_2 in the atmosphere, due mainly to the burning of fossil fuels could cause a warming trend leading to climate change in the next century. While there is still scientific uncertainty about the pace and impact of such a change it could be of sufficient magnitude to produce major physical, economic and social dislocations on a global scale. A special challenge posed by the issue is that if preventive actions are postponed until some effects are perceived, it may be too late to avoid appreciable climate change. It is therefore in the interests of all nations to support international efforts to understand the phenomenon at early enough stage for action to be taken to reduce or avoid major economic, social and environmental disruptions around the world. However, as developed countries are currently responsible for over two-thirds of total world emissions of CO_2 by fossil fuel combustion, they have a special responsibility for supporting and contributing to accelerated efforts to assess more precisely the nature of the problem, and the alternatives for action to reduce or avoid any serious risks.

Another issue is the world-wide distribution and use of harmful or potentially harmful chemicals. There are now some 70,000 chemicals on the commercial market, and many are currently used and released into the environment with little or no knowledge of their potential long-term effects. It is estimated that another 1,000 new chemicals enter the market each year. Chemicals cross national boundaries in a variety of ways; they are transported by winds and water.

(b) Risks to Ozone Layer

Ozone is a different form of life sustaining oxygen, having three atoms instead of the usual two. But this third atom, scientists have found, is very unstable and can be eliminated by chlorine ions.

The scientists have now found a close link between the family foam mattress, refrigerator and the air-conditioning system of an average home and Ozone Layer, 50 to 70 kilometres above the earth. They believe that free chlorine ions into the atmosphere, from plastics and other chemicals remove that extra atom of oxygen which makes Ozone. Only, in this case, the loss of the one atom changes the ability of the layer to act as strainer to absorb the damaging ultraviolet B wavelengths from the sunlight.

The findings by the experts have revealed that the rate of Ozone Layer depletion is sufficiently large to present serious consequences for man, animals and plants. In addition, new threats to the Ozone Layer are being recognized.

Take, for example, the refrigeration industry, food preservation has become a necessity and refrigeration is here to stay. Only advanced technology can provide the solution to prevent the release of the damaging ions. Soils are normally much better able to resist acidification than lakes, rivers and streams. Meanwhile, acid accelerate corrosion in most materials used in construction of historic buildings, and other important cultural objects. Some of the world's greatest cultural treasures including parthenon in Athens and Trojan's column in Rome are being eaten away by acid fall out.

There must be more effort to research and develop ways of improving energy conservation, environmentally appropriate technologies of producing heat and power, tend techniques for removing sulphur from fossil fuels and gaseous emissions.

The problem of acidification may well spread to new areas as a result of rapid industrialization and the growth of cities in other parts of the world, particularly developing countries. So, it is important that areas affected by acid deposition and susceptible too damage from acidification are identified as soon as possible. If they are, the damage could be mitigated or even avoided at a minimum cost to society by initiating research and applying what we already know about four pollution control.

(c) Pollution of World's Oceans

The data collected from the Baltic, the North Sea, and some North American and Australian estuaries and coastal waters suggest that in most of these areas the pollution that caused most concern was due to sewage, agricultural chemicals, oil and metals. Metal concentrations were clearly elevated in coastal waters, and in fish and selfish living there. In some areas, mercury levels in species such as fauna were high enough to make these fish unsuitable as human food. (Perhaps the most notorious incident of all is the outbreak of "Minamata Disease" in Japan. Methylmercury discharged from a chemical factory into the sea, or produced in the sea from inorganic mercury discharges, contaminated the fish eaten by local people at the town of Minamata of Kyushu island, Japan. As a result of this and a similar incident at Nilgata on the east coast of Honshu, nearly two thousand people have so far suffered from neurological disorders about hundred have died.

Some of the pollution came via river; the amount of iron, manganese, copper, zinc, lead, tin and antimony that reached the sea by this route was far greater than would be supplied by natural geological processes. Other contaminants came through atmospheric deposition: the importance of this pathway for metals and synthetic chemicals was increasingly recognized during the decade (1972-82).

Offshore oil and gas exploration and dredging for stand and gravel in coastal areas increased during the period 1972-82. Coastal zone development affected extensive estuarine areas, as well as mangrove swamps and coral reefs. Oil pollution killed sea-birds, fouled beaches and affected tourism. Although tanker accidents were the source of less than 5 per cent of all the oil entering the sea, accidents released large volumes in small areas, and were therefore especially damaging.

A recent report issued by a group of experts on marine pollution and published by the United Nations Environmental programme, warns that increasing contamination of the coastal seas, which has so far only caused localized disturbances to the marine environment, may become much more serious. The combined differs of the dumping of sewage sludge, industrial wastes, litter and oil spills into the Mediterranean, the North Sea, the Baltic and the Gulf of Mexico, may well cause ecological disasters which could gradually lead to a global crisis since the oceans are a unified life support system. The report points out that the sea's capacity to absorb wastes has its limits, so that as the world economy continues to grow, even more sewage, industrial wastes and other effluents are likely to find their way into the sea.

Man-made radioactive materials have been discovered in the open ocean and scientists have found traces of DOT in the Arctic sea and the Antarctic penguin, thousands of miles away from the areas where the pesticide is used. And because many of the disturbances which are introduced occur at vulnerable points such as the surface of the sea or on the sea floor, where oceanic life-supporting processes are generated, the dangers are magnified.

Although the modification of this microlayer has occurred only on the vicinity of oil spills and around the marine dump site of municipal wastes and no evidence has yet been found that detergents in the open ocean have significantly influenced the properties of the sea surface, the report points out that environmental catastrophies in certain coastal areas and seas might already have occurred. Much work however, is still to be done before the several ecological subsystems which constitute the ocean are fully understood. By the end of 1982 action plans were in operation or planned for 10 regions involving some 100 coastal and marine national parks and reserves were extended.

(d) Desertification

Desertification is now perceived as an intensifying worldwide threat which, to varying degrees, affects over 100 Member states of the United Nations. More than 20 per cent of the world's population (850 million people) live in the 35 per cent of the Earth's land surface (45 million km) which consists of the arid, semi-arid and sub-humid zone at risk of desertification. Three quarters of this area is already at least moderately desertified. Each year an additional 21 million hectares of agricultural land deteriorates through desertification to a point at which it is no longer economically productive. The last production has been valued at 36 billion dollars annually (1980 dollars).

Although desertification is exacerbate by several drought, its principal cause is human over-exploitation of dry lands through overcultivation, overgrazing, poor irrigation practices and deforestation. When drought strikes the overtaxed livelihood systems, desertification results.

The most devastating environmental and economic effects of desertification are in developing countries in the drylands. National balance of payments problems and local rural poverty lead to overexploitation of lands and forests for food, fuel, cash crops and meat for export. As these and other pressures increase to unsustainable levels, agricultural productivity and soil fertility decline, creating in turn new pressures which lead to further degradation. Recovery of land for agricultural purposes become costly and often economically unfeasible, thereby reducing the economic potential of the affected countries. In many of these countries agricultural land is also being reduced by salinization, urban sprawl and sanddune encroachment.

Increased desertification in one part of the world can indirectly induce desertification elsewhere, as loss of agricultural land, increased food demand raise prices and intensify pressures on marginal drylands in other food exporting and importing countries alike. There are also other impacts on international trade, to the extent that developing countries affected by rapid desertification will have a reduced capacity to import goods and services from developed countries and from other developing countries. Other globally shared impacts include loss of valuable genetic resources, disruption of hydrological cycles and increase of atmospheric dust.

A great deal is already known about the causes and effect of desertification and about the measures needed to half or combat it. Moreover, there is favourable cost benefit ratio to justify, preventive action. The 1980 estimate for implementing the 1977 plan of Action to combat desertification was 2.4 billion dollar annually over 20 years, which is only a fraction of the estimated value of agricultural production lost each year through desertification.

It is, therefore, in the interests of all countries to establish some concrete cooperative action between groups of developed and developing countries to half further desertification a goal which can be achieved, given a sufficient allocation of resources, on the basis of currently available knowledge.

(e) Loss of Tropical Forests

Although tropical forests are found in early half of all United Nations Member States, most are in twenty countries and nearly half their total area in three countries alone (Brazil, Indonesia and Zaire). The tropical forests are the world's richest biological zones, and are estimated to contain as much as 40 per cent of all terrestrial species on the planet. In addition, tropical forests provide a wide range of useful products of economic significance for both developing and developed countries. Undisturbed tropical forests are also home to millions of the world's tribal peoples.

Many expert contend that tropical forests are being exploited at a rate and in a way that is ecologically destructive and economically unsustainable. There are varying estimates of how much of the tropical forests will be lost by the year 2000 through complete conversion to other uses and buy severe degradation. The results of the tropical forests resources assessment project conducted by FAO and UNEP indicate an estimated minimum loss of closed tropical forests of 125 per cent by the year 2000 (an average of 75 million hectares annually between 1980-2000) though complete conversion to other uses alone. Other studies which not only consider complete conversion but also include severe degradation caused, for example, by logging operations, by overgrazing, by overexploitation for fuelwood or by natural causes such as fires, insects and diseases, have estimated losses of tropical forests as upto 20 million hectares or more a year. The FAO/ UNEP study estimates that tree planting on industrial and non-industrial plantations accounts for only 10 per cent of the area of tropical forests cleared annually.

The major causes of deforestation differ by region. According to the FAO/UNEP report, in Latin America the two major causes are shifting cultivation and the conversion to pasture land for cattle ranching. In Asia and Africa, shifting cultivation by subsistence farmers and removals for fuelwood are major causes. The extraction of timber from tropical forests for industrial purposes is not considered by the FAO/UNEP reports as a major factor in deforestation. On a worldwide basis, the FAO/UNEP report estimates that fuelwood and charcoal removals from tropical forests are as much as eight times greater than industrial wood.

Serious degradation of loss of watershed forests in particular has a wide range of major ecological and economic effects through increased erosion, floods landslides, more silting of hydro electric facilities, irrigation systems, reservoirs and harbours. The livelihood of half the world's population depend directly on the wise management of watershed forests and eco-systems.

Deforestation also influences local, regional and possibly even global climatic conditions by altering hydrological regions, reducing vegetation cover and causing more heat and light to be reflected back to the atmosphere (the albedo effect), and through accelerated burning of the forest biomass, which contributes to carbon dioxide build-up in the atmosphere and the global warming trend. The economic, social and political implications of possible regional and global climate change have only began to be assessed.

Though the main causes immediate effects of deforestation are in the developing countries, there are significant environmental and economic effects which are of increasing concern for developed countries also. Major concerns include the availability of genetic resources for agricultural improvements, for medical research, and for the pharmaceutical and biotechnology industries; and the possible impact of deforestation on global climatic conditions.

(f) Soil Loss

Almost every environmental concern connects clearly with the importance of soil. Soil acts as a reservoir for water and other inorganic substances cycling through the eco-system.

It provides habitat for many of the different populations of decomposer organisms that control the rate and capacity of nutrient recycling.

The importance of soil to agriculture can hardly be over emphasised. Agriculture provides about 44 per cent of the poorest country's gross domestic product and 83 per cent of their employment.

Growing more food, building irrigation projects, conserving nature and precious wildlife and other needs must depend on proper soil management. The wideranging importance of soil confirms that soil management must consequently be many-sided. The broad range of natural and man-made stress that cause the two type of soil damage: physical displacement and alterations in solid structure and composition.

Uncontrolled shifting cultivation, salinization and waterlogging flooding and plant cover destruction are same. An appropriate use of chemicals, weapons testing and deployment in warfare are some other causes of soil stresses. Together or individually they wield a destructive force that is manifest in soil erosion, the loss of soil nutrients and other types of soil degradation.

Soil damage is also caused by patterns of dry land mountain side cultivation that are carried out without arrangements for bonding and terracing to protect the topsoil.

Continuous cropping to meet growing food needs leave little time for soil conservation techniques.

Deterioration of the productive capacity of the soil is most acute in areas affected by erosion, drought, mineral stress and other elated factors. Soil resources that are vulnerable to these factors constitute 82 per cent of all lands in South Asia.

While erosion is an integral part of the natural system, occurring even when land is covered by grass, or trees, its extent and intensity can be minimized provided land denudation is correspondingly reduced. The fact that it takes nature 100 years or more to build one inch of top soil should have engineered the vigilant observance of soil conservation measures. Instead, problems of soil erosion have continued ticking.

Countries in East and South Asia are particularly burdened with soil erosion occurring in watersheds.

Anticipated acceleration of agricultural production and other activities between now and 2000 is casting its shadows: existing cultivable land will be further subjected to intensified use as cultivation of unsuitable areas persists: logging, firewood gathering and grazing will accelerate the elimination of vegetative cover, while the construction of roads, mine and water resource projects will be further pursued in the course of development. Such prospects place the least developed countries in an even more precarious situation.

Slash and bum farming is a common enough culprit. Wielding destruction in gradual succession, it initially cause the loss of fertility by the thin topsoil, and later eroding its stability. In India, slash and bum farming is particularly rampant in the north-eastern and central regions, affecting between six and eight million hectares of land.

Salinization affects 80 million hectares (8 per cent) of total in South Asia.

Something should also be said about agricultural techniques which are designed to raise land productivity but instead lead topsoil loss. The use of cheap nitrogen, for example, replaces nitrogens, fixing legumes, resulting in loss to topsoil.

The commonality of problems, in addition to the implications of transboundary soil management points up to the fact that soil problems cannot always be resolved within the context of national concerns. Clearly, what are required are mutual policies and strategies concerning land management, and the World Soils Policy provides an appropriate starting point.

Working in collaboration with the relevant UN bodies, UNEP has proposed the World Soils Policy programme which endeavours to enhance international awareness, assist countries in the formulation of national soil policies, help to develop technical and scientific knowledge and collect, compile and disseminate data on the use and management of the world's soil resources.

(g) Depletion of genetic resources

25,000 species and more than 1,000 vertebrate species and subspecies are now estimated to be threatened with extinction, yet these two categories of organisms represent only a small proportion of the terrestrial species on the planet. According to some estimates as much as from half a million to a million species would be extinguished over the next two decades. Extinction of species on this compressed scale would be without precedent in human history.

As much as two-thirds of all terrestrial species and the majority of endangered species are located in developing countries. A significant loss of species in developing countries would have a worldwide effect on future agriculture, industry and science, and on human health and welfare generally. The most serious threat to many species is the disruption and destruction of their habitat by man, especially in the tropical forests which alone harbour as much as 40 per cent of all terrestrial species.

Significant species loss could directly effect human health and welfare locally, and even nationally, in many developing countries, where wild plants and animals are a source of food or traditional medicines and income, especially for the poor in rural areas. In some countries tourism largely on wildlife is a major source of foreign exchange.

In many developed countries over one-third of the gain in the productivity of many major agricultural crops over the last few decades has been attributed to breeding programmes using genetic material from related wild plant varieties, many of which are found in developing countries.

Moreover, many of the major crops in developed countries now have a limited genetic base. For example, more than half the Canadian prairie, wheatlands are devoted to a single variety, Neepawa, which was developed with germplasm from Kenya, 72 per cent of United States potato production depends on only four varieties; and the entire United States soyabean industry is derived from a mere six increasing the yields of major crops, such as wheat, maize, rice, potatoes and for combating new strains of diseases or pests which threaten them.

Genetic stocks from one part of the world frequently make an enormous contribution to another. Current research shows that enormous opportunities lie ahead, such as the utilization of Asian blue-green algae to fix nitrogen for fertilization, which may constitute a gain of hundreds of millions of dollars a year to the economics of developed countries alone.

Major medicine and other pharmaceutical products in developed countries are also often dependent on plant and animal's species. For the United States of America it has been estimated that more than 40 per cent of the prescriptions written each year contain a drug as natural origin-from higher plants (25 per cent), microbes (13 per cent) or animals (3 per cent)-as the sole active ingredient or as one of the main ones. It has been estimated that among the 10 per cent of species that might be extinguished by the year 2000 there could be many plants with potential for use in new plant based drugs and pharmaceuticals with a combined worldwide commercial sales value of as much as 200 billion pounds annually.

Biomedical research in developed countries is also vitally dependent on wild species, many of which are threatened with extinction. The innovative capacity and future of the rapidly expanding field of biotechnology, practised largely in developed countries, relies considerably on genetic patterns and resources from many wild plants and animals. As only about 10 per cent of the world's plant species and 1 per cent of the animals species have so far been subjected to even a preliminary screening for potential medical or commercial applications, a significant loss of species could also represent a permanent loss of sources for more effective chemicals, medicines and other useful products.

In spite of this interdependent relationship, species continues to accelerate. Developing countries often lack the capacity and economic incentive for extensive species protection. They have derived little or no economic benefit from the development and use of genetic materials in developed countries, where nevertheless, developing countries are and will remain the principal source of genetic materials from wild plants and animals.

This clearly shows that there is a common incentive for developed and developing nations to prevent a potentially catastrophic loss of species. The mutual advantage of comprehensive joint effort could result in new medicines and other useful product that would contribute to improved levels of human health and welfare in all countries.

Relationship Between Environment And Development

At the time of Stockholm Conference, developing countries were hesitant about environmentalism. Seeing it as potential restriction on development. It has now been demonstrated that "environment" and "development" are compatible and that to ensure lasting improvement in the quality of life, development must be environmentally sound and sustainable.

Poverty is by far the most important cause of environmental destruction. It degrades the human environment and in doing so obstructs development. For instance, the lack of safe drinking water for over 1300 million people and of sanitation for over 1700 million of the most severe form of pollution caused by poverty-responsibly, according to WHW, for 25000 deaths a day and for 80 per cent of world disease. The death, debilitation and diseases caused by contaminated water have been a severe brake on development: a survey carried out by a South American Government showed that the provision of clean rural water can recoup coasts five to seven times over by the number of working days it continue to save.

The many other ways in which poverty contributes to the cycle of destruction are equally far reaching. Lack of adequate shelter increases vulnerability to natural disasters, and the marginal settlements in the rapidly expanding urban areas of the developing world are prime examples of degraded environments. Desertification, deforestation, soil erosion and salinization, result as inappropriate agricultural practices are used to meet the desperate need for food.

No development which does not tackle the main causative agent of environmental impoverishment worthy of the name.

Over-consumption and wasteful use of resources by the developed nations and the privileged strata in the developing world pose a threat to the environment comparable with poverty. The build-up of carbon dioxide in the atmosphere and industrial pollution are only a few manifestations among many. However, the global problems created by inequitable development go far deeper. The demands of the developed countries for resources from the developing nations add to the pressures on their environment. The commitment of increasing amounts of the most fertile land in the developing countries to grow cash crops for consumption in the industrial countries provides an instance. Unsustainable demands of fish resources and tropical forest cover are two more.

Thus, the effect of pollution increase and people's consumption and production has been to upset the balance between people and resources, hence, leading to deterioration of the environment. The economic future of developed and developing countries is not dependent only on the laws of trade and finance but on environmentally sound development cooperation. To be beneficial, such co-operation should deal not only with the relations among men, but also with the relations between man and nature.

Agricultural development in the developing countries should not be seen exclusively as a basic economic activity which provides the food for human sustenance, or as an activity to provide employment for growing populations, but should be considered from the point of view of maintaining sustainable production. In appropriate methods of land cultivations as well as backward land tenure systems have not only produced less than optimal yields, but have also contributed to increasing land exhaustion, as well as to salinization, erosion and desertification. Increased use of fertilizers and pesticides combined with large scale irrigation will greatly affect the environment, and care is needed if the overall impact is to be positive. Efforts directed towards modernizing world agriculture should take fully into account the environmental connections of this basic human activity, so that the objective of producing food and providing employment may be reached in the short-term, and the achievement sustained in the long term.

The maintenance of genetic resources and diversity is an important environmental issue, but it is seldom realized just how important it is to economic development. The rapid shift to intensive agriculture and monocultures had created a vulnerability that requires the maintenance of natural stocks. The economic value of these genetic reserves-located for the most part in the developing world-is difficult to quantify in strict monetary terms, but certainly it amounts to billions of dollars each year.

The dependence of the world economy on petroleum, a non-renewable source of energy, become clear in the last decade of rapidly rising oil prices, as did the importance of conservation and of diversification and substitution in sources of energy. As regards diversification and substitution, there is increased reliance on coal and on nuclear energy, with new actual and potential impacts on the environment. Clearly, any energy transition must be achieved in full knowledge of the environmental impacts of various forms of energy production and use, and must be based on rational decisions on the desired energy mix. Assisting developing countries in making the transition successfully must be seen as serving the interests of the global environment.

Major environmental problems also occur because of the unchecked use of often environmentally imprudent technologies, or of the use of technology in a manner world-wide inappropriate in the light of current knowledge. Examples of such problems are toxic wastes, threats to the ozone layer and to coastal areas, effluents from feeding lots, possible climate change, etc. The economic effect of such environmental problems could be severe. Indeed, technology is crucial in terms of the impact, positive and negative, its use has on the environment. In the industrialized countries, the capacity to as certain and evaluate these impacts in growing. However, only a small percentage of the world's scientific and research capacity is found in the developing countries. The current pattern of technology use as therefore, been elaborate in the economic, social and environmental conditions of the developed countries. The environmental soundness of any particular technology in the local conditions for which it is intended should be a fundamental element in deciding upto its appropriateness.

Developing countries are facing increasing difficulty in their efforts to find adequate resources, particularly foreign exchange, to step up their pace of development. This has often compelled them to focus on shorter-term aspects without giving adequate weight to longer-term considerations. The high real interest rates prevailing in the last few years have tended to reinforce these effects, since high interest rates tend to skew cost/benefit assessments unduly towards short-term benefits. Environmental issues, the impact of which is felt over longer period, are discounted by this forced short-term perspective. A one per cent increase in interest rates adds approximately 5 billion dollars on the debt burden of developing countries. To have increased its export earnings (not profits) by one billion dollars in 1981, South American as a whole would have had to increase its banana exports three-fold, Ecquador five-fold and Colombia nine-fold, while leading cotton exporters like Egypt and Turkey would have had to double and triple their cotton exports. This would have meant bringing million of additional hectares of land into production to grow these export corps.

The world is undergoing a period of technological evolution owing, among other things to a growing understanding of the energy and resource use and environmental impacts of current technologies. Thus, new technologies, developed with such understanding, tend to be environmentally more being than those they supplanted. The extend to which developed countries apply such new technologies is affected on the one hand by their commitment to existing industrial patterns and on the other by their greater economic capacity to invest in them: developing countries, on the other hand, are less committed to existing industrial patterns, but take the economic capacity to invest in new technologies or to adequately evaluate technologies. For developing countries to be tied environmentally inappropriate technologies is ultimately counter-productive not only for their environment, but also for their economies. Increased transfer of technology to developing countries should be accompanied by the provision of data which would permit informed choice in the light of local conditions, and the capacity of these countries to evolve suitable technologies should be strengthened.

The central question today is not whether to choose between development/industrialization and environment. It is how to select patterns of development that not only minimize adverse impacts, but also are actually designed to stabilize and improve environmental, and hence economic conditions.

Environmental Problems Faced By Developing Countries

Following are the major environmental problems faced by the developing countries:

(i) The most important concern relates to the degradation of land and water on account of the very extensive deforestation, denudation, soil erosion, waterlogging and salinisation, desertification, floods and drought etc. These are the matters which constitute a threat to the very survival, let along the well being of a large number of countries which are already beset with the most serious problems of poverty disease and growing populations. In fact, the countries which are affected with serious problems of land degradation are finding it difficult to secure sufficient resources even to fight on this front, so urgent and unending is the demand for funds required for the relief of population repeatedly hit by floods, droughts and rural unemployment caused by land degradation itself.

(ii) The preservation of flora and fauna is threatened by deforestation and denudation in most developing countries. It must be saved from extinction in our own long-term interests.

(iii) Hazards to human and ecological health posed by various kinds of pollution whether chemical, radio activity, or thermal are well known. What is however, not adequately appreciated is that as far as developing countries are concerned, the pollution caused by insanitary conditions represents a far greater health hazard to impoverished populations which are already suffering from malnutrition and debility. The disease and ill-health attributable to the lack of facilities for the proper handling disposal and treatment of human animal and domestic wastes are compounded by the contamination of drinking water supplies. It is not, therefore, at all surprising that the UN should have declared the present decade to be the International Drinking Water Supply and Sanitation Decade.

5

POLLUTION CONTROL

Historical Perspective

Despite the Factories Act, 1948, and the licensing procedures framed under the industries (Development and Regulation) Act, 1951, containing provisions for regulating the setting of industrial unit from the point of view of environmental hazards, especially for the industries which emit poisonous and highly toxic gases endeavours at environmental protection-atmospheric as well as water did not make much head way till about the middle of the last decade. An Expert Committee of ASSOCHAM, set up in 1973 specifically to asses the situation, reported after the survey, that there was definite lack of awareness of the environmental protection problems even among the constituents of ASSOCHAM. The position was no different, if not worse, in the case of the industrial sector in general. Most industrial unit did not take any environmental protection measures. They even lacked monitoring facilities in this regard.

The Committee therefore recommended that adequate awareness of the magnitude of the issue of environmental protection has to come about in time so that its cumulative pressures could be abated it, however, felt that the government would need to be considerate to the extent that a frank exposure of industry to the problem was not inhibited at the start. Legislative measures, it pointed out, could only impose penalties. But so long as industrial management did not themselves face upto the problem as a menace. Its abatement was unlikely to be achieved. The Committee stressed that environmental control, indeed, was management problems. It was, therefore, imperative that a proper awareness of its existed at the appropriate levels of management.

The reasons for this inadequate appreciation of environmental protection in the Indian industrial sector as reflected by the 1973 survey of ASSOCHAM, of course, are not far to seek. The industrialisation of the country till about the early 1970s had not reached a stage where environmental hazards could be said to have gained managing proportions. There was more concern over ecological and environmental degradation caused by deforestation, soil erosion and such other factors; which led to what has come to be known as the "Chipko Movement" i.e., save trees and forests. Efforts has also been initiated to contain the Rajasthan desert.

Environmental problems caused by domestic and industrial wastes, were a matter of concern primarily in a few metropolitan cities where density of population, vehicular traffic and industrial units was acute. Further, owing to the fact that industrial development in the vast tracts elsewhere in the country was just in the nascent stages; industrial units were being located mainly on the basis of raw material availability, access to the market, transport facilities and such other techno-economic considerations, without adequate attention to environmental aspects.

Interestingly enough, concerted endeavours at containing environmental hazards seemed to be lacking till then even in many industrialised countries, despite their concern over acidic rain an eloquent testimony of this is available in the fact that the industrialised countries began to analyse the problem seriously only after the 1972 Stockholm Conference of the United Nations on Human Environmental in which some 113 countries participated, this UN Conference promoted the International Chamber of Commerce to draw up guidelines for industry for environmental protection, which came to be published for the first time in 1974. These were revised in 1981. Another review of them has been taken up recently (Annexure-1), clearly indicating that the seriousness of the problem has come to be appreciated fully only in the past few years.

Incidentally, the International Labour Organisation, too, started taking keen interest in environmental protection only after 1972.

Indian industry, thus, was not an exception in regard to appreciation of the problem of environmental protection till the early 1970s. The industrial safety aspect within the premises of industrial units, of course, had all along been taken care of, by and large, by industrial managements on the lines prescribed by the Factories Act and Industrial relation legislation. The Indian Labour Laws, in fact, has been fully conforming to the guidelines laid down by the ILO. In several respect they were even more progressive than in many advanced countries. The position continues to be so.

The situation on the environmental front, however, has fast undergone a change in the recent past. The pull towards urban conglomerations in search of employment, and consequently their rapid growth, has created environmental hazards on an increasing scale. The establishment of many high-technology industries also has added new dimensions to the problem not withstanding the fact that the large and medium-scale units in them have been aware of the hazards involved and, by and large, have been taking the necessary precautionary measures. The proliferation of small units in such areas of industrial activity, of course, has accentuated-the problem to some extent.

A country seeking increasing presence in high-technology areas of industrial activity, it is well recognised, cannot turn a Nelson's eye towards environmental protection. The pace of industrial development envisaged currently, it is further felt, warrants that environmental control ought to be accorded high priority lest the situation should aggravate further and become difficult to tackle, technically as well as financially. The establishment of the Department of Environmental at the centre in November, 1980, and the setting up of the Central Ganga Authority recently to cleanse this sacred river of pollutants are testimonies to the growing awareness on the part of the government of the imperative of environmental protection.

Statutory Provisions

Relating to Industries

- The State Director of Industries confirms that the site of the project has been approved from the environmental angle by the competent state authority.
- The entrepreneur commits, both to the State Government and the Central Government, that he will install the appropriate equipment and implement the prescribed measures for the prevention and control of pollution.
- The concerned State Board of the Prevention and Control of Pollution has certified that the proposal meets with environmental requirements and that the equipments installed or proposed to be installed are adequate and appropriate to the requirements.
- The guidelines issued by the State Directorates of Industries provide information relating to the areas to be avoided for the sitting of industries, precautionary measures to be taken for site selection as also the aspects of environmental protection to

be incorporated during the implementation of industrial projects.

Notified Polluting Industries

Twenty highly polluting industries have been notified by the Central Government as falling within the purview of industrial licensing. These are:

1. Primary metallurgical industries, viz., zinc, lead, copper, aluminium and steel;
2. Paper, pulp and newsprint;
3. Pesticides/insecticides;
4. Refineries;
5. Fertilisers;
6. Paints;
7. Dyes;
8. Leather tannin;
9. Rayon;
10. Sodium/Potassium cyanide;
11. Basic drugs;
12. Foundry;
13. Storage batteries (lead acid type);
14. Acids/alkalies;
15. Plastics;
16. Rubber-synthetic;
17. Cement;
18. Asbestos;
19. Fermentation industry,
20. Electro-plating industry.

An environment specific clearance is necessary from the Department of Environment as a pre-condition for setting up an industry unit in these lines of manufacture. Installation of equipment prevention and control of pollution is insisted upon in such industrial units before the commencement of production.

Environmental Protection Enactments

In respect of water and air, the two major subjects of environmental concern, specific legislation currently in force are-The Water (Prevention and Control of Pollution) Act, 1974, and The Air (Prevention and Control of Pollution) Act, 1981. Under the former act, the Central Government and most States have constituted Central and State Boards for the prevention and control of water pollution. They have been entrusted with the tasks relating to the latter act as well. Land pollution is also covered by the former act.

In 1977, another legislation was put on the statute. The Water (Prevention and Control of Pollution) Cess Act. Its purpose is to augment the resources of the Central and State (Prevention and Control of Pollution) Boards through the levy of a cess on the consumption of water by specific industries and local bodies at rates specified in the schedule to the act. The amount so collected is made available to the State Boards according to a prescribed formula for being utilised by them to carry out their functions. The Central Board is the apex body attending to the problem of water and air pollution at the national level and advises the Central Government in formulating all the policies and plans for action of environmental pollution. It acts as the State Boards in the case of some Union Territories and coordinates the activities of the State Boards in the nation-wide implementation of pollution control and monitoring through the regional offices in the east and north-east at Calcutta, in the north-west at Chandigarh and in the south at Hyderabad. The States in the north-central region are looked after by the Central Board Head Office at New Delhi. A regional office for the western zone is being set up.

New Developments

The Minimal National Standards (MINS) for pollution discharge for specific industries have been formulated and control measures are being implemented.

In the wake of the Bhopal tragedy, caused by the leakage of poisonous gas in a pesticides manufacturing plant there, a policy paper for legislative and institutional framework to regulate the import, manufacture, handling and disposal of hazardous substances (toxic chemicals and microorganisms) has been prepared. A document recommending the Threshold Limit Values (TLV) for selected hazardous industries for environmental purpose also has been prepared.

Water Pollution Control Status

The Central and the State (Prevention and Control of Pollution) Boards last year completed a country-wide inventory of water polluting large and medium industries, long with the effluent treatment plants status, as the end of December, 1984, it was released recently.

This inventory points out the 4,054 industrial units in the country, 1,731 large and 2,323 medium sized-which had been identified as relevant to water pollution, 51 per cent (2,075) units in the large-scale sector and 39 per cent units in the medium-scale sector had adopted appropriate pollution control measures.

The State of Karnataka, Gujarat and Maharashtra led in the setting up of effluent treatment facilities with 95 per cent, 86 per cent and 80 per cent of the units there, respectively, having such facilities. The other states which had been identified as important from this view point but had lagged behind were Andhra Pradesh, West Bengal, Tamil Nadu, Haryana and Uttar Pradesh, seventy-six per cent of the 4,054 industrial units identified as relevant to water pollution were situated in these eight states.

Industry-wise Status

The industry-wise survey in the inventory suggests that a significant improvement in the situation had taken place between April, 1981 and December, 1991, in the setting up of waste water treatment facilities in the eight priority category industries included.

In the implementation phase-I (Sixth Plant Period). These were sugar, distillery caustic soda, fertilisers, oil refining, man-made fibres, iron and steel and textiles. While the growth in the number of units in these industries during the period of survey had been 48 per cent, the number of units having waste water treatment facilities had gone up by 60 per cent. Overall, 51.4 per cent units had effluent treatment facilities at the end of 1984, as against 47.4 per cent in April, 1981.

The industries other than the above eight priority category industries are to be covered for water pollution control in the phase-II of the programme during the Seventh Plan period. The progress between April, 1981 and December, 1984, in their case had been even better, as 31 per cent of the units in these industries had effluent treatment facilities at the end of 1984, compared to 36 per cent in April, 1981. Their total numbers grew during this period of 236 per cent and the number of units with effluent treatment facilities went up by 79 per cent as per the claim of Government.

Industry-wise, among those industries in which there were 100 or more units having effluent treatment facilities were: pharmaceuticals (74); distillery (65); pulp and paper (62); and sugar (57).

General engineering, textiles, foodstuffs and edible oils and vanaspati units in this category of industries were lagging behind within percentage below 50 indicating insufficient progress. Among the industries in which the total number of units was less than 100, the units with waste water treatment facilities, percentage-wise, were as follows; oil drilling and refining (94); caustic soda (77); paints and dyestuffs (75); man-made fibres (69); insecticides and pesticides (69); fertilisers (67); mining (66); dairy (59); and thermal power (54). Iron and steel, cement, rubber and rubber products and mineral industries were yet to attain a satisfactory performance level among the category of industries.

The inventory, very significantly, points out that the progress in regards to pollution control among the eight priority industries taken up in the first phase during the Sixth Plan period was slower than in the second priority industries (to be taken in phase-II during the Seventh Plan) because the former included sugar, textile and distillery industries which are comparatively old, earn less profits and need higher investment for pollution control.

Fiscal Incentives for Pollution Control

With a view to encouraging industrial units to take to pollution control and fostering conservation and environmental protection. The following incentive are provided to individuals corporate bodies:

(a) Any sum paid by a tax-payer to any association which has as its object the undertaking of programme of conservation of natural resource to be used for such purpose, is allowed as deduction in computing his taxable income provided that the association/institution, as also the programme of conservation of natural resources, has been approved by the prescribed authority.

(b) Depreciation is allowed at the higher rate of 30 per cent, as against the general rate of 15 per cent, in respect of specified machinery and plant installed for control of air pollution, water pollution and solid waste.

(c) Investment allowance is granted at the higher rate of 30 per cent, as against the normal rate of 25 per cent, of the actual cost of new machinery and plant installed which would assist in control of pollution and protection of environmental and which have been notified in this behalf by the Central Government.

(d) When a company owning an industrial undertaking situated in an urban area shifts, with the prior approval of the Central Board of Direct Taxes, to another area, it is granted at a credit certificate with reference to the amount of tax payable by it on the capital gain arising formative transfer of capital assets being machinery and plant, lands and buildings.

(e) If any person, or local authority, liable to pay the cess under the Water (Prevention and Control and Pollution) Cess Act, 1977, instals any plant for the treatment of sewage or trade effluent, such person or authority is entitled to rebate of 70 per cent of the actual cess.

Amendments Proposed to Pollution Control Legislation

The Central Government has proposed to amend during the ensuring Budget Session of Parliament, in a comprehensive manner, the environmental protection legislation with a view to making it more effective and deterrent.

The amendment to the Water (Prevention and Control of Pollution) Act is reported to be provided, that the onus of proof in the case of prosecution would rest on the accused parties. They would have to prove that they were not guilty of charges. At present, the prosecuting agencies—the Central and State Boards for the Prevention and Control of Water Pollution—have the responsibility to prove the charges in the court.

The amending bill is also said to be proposing to enabling Courts to give 'ex-party' decisions. It is also proposed to amend the Factories Act with a view to protecting workers and the people living near industrial units against health hazards arising out of storage, use, production and Disposal of hazardous materials. This act is also envisaged the permissible limits of exposure to toxic chemicals and substances. Deterrent penalties are proposed for those who violate the regulations.

Minimal national standards have been finalised and prescribed by the Department of Environment for clearance of locations for new units in the clor-alkali sugar, man-made fibres, textiles, oil refining, distillery and fertiliser industries. Similar relation to pulp and paper, pharmaceuticals, pesticides, thermal power plants and organic chemical industries is expected to be brought out shortly. Standards for small-scale paper plants and non-ferrous metals industries are stated to be under preparation.

Environmental Protection, thus, is receiving serious attention in the highest echelons of administration. It is also hotly debated subject in industrial circles. Not only have the two apex industry and trade organisation in the country, as well as the other specialised agencies, set up expert. Committees to deliberate upon the issues involved, but their constituent, too, have been very active in following suit. Quite a few seminars/workshops have been organised already to discuss the problem and throw up ideas to tackle it effectively. Industry-wise deliberations, too, have taken place in several cases. A brief resume of the trend of these deliberations will be useful for framing the issues for this workshop.

Standard Setting

Industry, it may be stressed, of course, is not averse to taking energetic measures to reduce environmental hazards to the minimum. Reference has already been made to the improvement that has taken place in containing water pollution caused by industrial units. What the industry desires is that the environmental protection norms and measures should be drawn up realistically so that they implemented effectively. Obviously, there is no sense in setting up norms which cannot be met within a specific time-frame for reasons that can be anticipated.

The major problem confronting the country, in so far as environmental protection is concerned, does not so much relate to new industrial units as to the old ones. The environmental angle is well-taken care of in the case of new units. The problem in their case arises from such factors as the latest machinery and equipment not being installed. Sometimes, allowing congestion in their neighbourhood, inadequate attention paid by some management to preventive maintenance and their financial involvements.

Constraints

Old industrial units, however, suffer from several structural disadvantage and are not, therefore, always able to put into effect even some of the required measures. They are mostly based on technologies which when developed were not much concerned with environmental protection. Dove-tailing modern equipment with the presently installed machineries invariably poses practical utilities. Space limitations, too, are a major constrain up pollution control equipment in a large number of even if they co-operate fully is going in for these. This is particularly the case with the units in the cities.

Then, these are financial constraints. The financial viability of most old units has been shaped by rigorous price controls, formal or informal, to which they have been subjected over long periods.

The closure of some units, apparently, is not a practical proposition, nor it is desirable as a fairly large number of them can well be revived through modernisation. But even when a sick industrial unit cannot be revived, it invariably is not allowed to close down for some reasons or the other. The shifting of such units from congested areas is also hindered by the local authorities or workmen. The Union Government, it may be recalled, introduced a provision in the Income Tax Act nearly two decade ago, providing for tax credit certificates for shifting of industrial units from urban areas-of course, on obtaining specific approval for the purpose. How many units were allowed to take advantage of this measure? Very few!

Differential Treatment for Old Units

The old units, with technology which did not take into account environmental protection, obviously, need to be treated on a different footing. The environmental protection norms in their case cannot be fixed as strigently as in the case of the modern projects the equipments for which invariably take into account the need for environmental protection.

Even in the case of new units, environmental norms ought to take into consideration the equipment manufacturing technology currently available in the country. Self-reliance in equipment supplies is one of the corner-stones of our industrial policy, imports are allowed either when the technology involved is not available within the country or it is high-tech. Mis-match between environmental protection norms and the available equipment supplies, apparently, needs to be obviated.

Locational Consideration

There are two other important aspects of norms fixation. Firstly, the norms pertaining to thinly industrised areas in the beginning need not be as stringent as for locations where industrial growth has already taken place to a significant extent. Apart from containing production cost-environmental protection, indeed, means a good deal of investment which increases costs of manufacture the reasons for resorting to this course in such areas are: (i) absence of industrial culture, and (ii) thinness of pollution. In the industrially congested areas, mild norms will accentuate the already polluted environment. Such a distinction is made even in the industrially advanced countries. The norms can be progressively upgraded as workers develop industrial culture and get attend to sophistication in operations.

In the case of the highly polluting industries, however, the dissentiation in the norms need not be any large. In such cases, be better to have stringent norms to obviate a rapid deterioration in environment and large expenditure later on.

The above suggestions that while national norms for environmental protection ought to be fixed realistically, preferably by an expert institution such as the Indian Standards Institution, in constitution with industry, so as to make them practicable for effective implementation. The State Governments can be allowed some latitude to modify them in the light of the circumstances prevalent in their jurisdiction, particularly, location-wise. This will help in fostering industrialisation in the initial stages.

Phasing out of Implementation

Secondly, the older units, as mentioned earlier, not only suffer from some structural disadvantages in implementing environmental protection on measures but also face financial stringency in undertaking this task. Environmental protection in their case, therefore, has to be phased out over a reasonable period of time, besides providing them financial assistance at concessional rates.

Fiscal Concessions

The fiscal provisions for environmental control in some cases appear to be inadequate. To contain the costs of installing environmental protection equipment, it will be as well worthwhile to allow imports of the requisite equipment under open general licence, with exemption from custom duties. Locally produced equipment can be exempted from excise levies, sales tax and octroi duties, besides beings provided the benefit of international price parity in regard to the raw materials going into their manufacture. Imports of pollution control equipment in fact, ought to be treated on the same footing as imports for R & D endeavours and energy saving.

A very useful idea that can be considered for encouraging environmental protection is making available of a certain percentage of excise duties paid by particular industries instead of giving concessions under the Income Tax Act of write-off facilities for the equipment used for this specific purpose. The experience with this level in raising industrial output in the past has been often very wholesome.

Consent Feel, Cess

Another important point concerning fiscal measure relates to the consent fee charged by the State (Prevention and Control of Pollution) Boards from different types of industry under Section 21, sub-section 2, of the Air (Prevention and Control of Pollution) Act and the cess leviable under the Water (Prevention and Control of Pollution) Act. It has been observed that the State Boards collecting three to four times the prescribed current fees, considering each that of an industrial unit as an independent industry. This practice needs to be stopped as the judgement whether a particular plant is a separate industry differs from person to person.

As regards the cess, there appears to be no rationale for imposing it on the sea water used for cooling purposes. It amounts to several lakhs of rupees per month, even if levied at a concessional rate, in the case of coastal power stations. This water is neither consumed nor polluted.

The cess and the consent fee collected under the pollution prevention legislation, it may further be stressed, ought to be utilised for countering pollution of fostering research and development for the purpose than being considered as revenue for being utilised for some administrative purposes not connected with environmental protection.

Product Pricing

Environmental protection measures, though very essential, do entail extra expenditure which finds reflection in higher production costs. A natural corollary to this expenditure is that the additional cost should be taken into consideration fully in the cost appraisals of industries the prices of whose products are regulated.

Specific Industry Problems

There are experienced, sometimes, various practically difficulties in the implementation of environmental protection measures. In the case of distilleries, for instance, the pollution control equipment available elsewhere cannot be used in India because of the raw materials fed into the plants and the by-products generated. Some distilleries and equipment manufacturers have put up pilot plants involving advanced technologies, but their utility cannot be established. Also distilleries in India are of environmental makes the requirement of environmental protection equipment vary from plant to plant.

Power Water Constraints

The kinds of power and water supplies available from public utilities often make it difficult to set up sophisticated environmental protection equipment. Owing to frequent power break down, load shedding, low voltage, etc., not only do the pollution control measures fail, quite often, but also the levels of pollution go up due to the pollution control measures fail quite often, but also the levels of pollution go up due to the various processes involved. Water supplied to industries, sometimes does not confirm to the tolerance limits laid down, making it impossible for industrial units to control pollution at their level.

The progressive deterioration in the quality of coal supplied has aggraved the problem. Proper supportive measures for, from governmental agencies, including municipalities, for the adoption of pollution control measures quickly are often found wanting. Stringent standards, therefore, cannot be met by industrial units owing to factors their control.

Workman's Role

The role of workforce in containing environmental hazards cannot be minimised. Any lasity on the part of workers can play havoc, even if management is fully conscious of the need for environmental protect. The increasing interest being taken by the ILO in environmental affairs reiterates mis.

Management, of course, has significant responsibility for educating it workforce in pollution control measures. It has to educate its employee in this regard through induction training programme for the new entrants and regular inhouse courses for both the new entrants and the old employees.

In each hazardous unit, a step-by-step check-list of safety measures for each hazardous process and operation needs to be prepared and prominently displayed so that the necessary action can be taken by the shop-floor personnel in case of emergency. The latest information on hazards and safety measures ought to be collected and disseminated to workers directly or through their unions. Selected employees may as well be exposed to knowledge about industrial arranging participation for them in national and even international training courses. Periodical rehearsals through induced emergencies too, ought to be arranged.

Environmental safety, in fact, needs to be introduced in the educational system itself from the stage of schooling to IIls so that employers as well as employees get attuned to this necessity from the very beginning.

Pollution Monitoring

The another important for undertaking environmental protection measures in a systematic manner, the need for proper monitoring of environment. It is rather unfortunate that environmental monitoring and testing facilities have left much to be desired, notwithstanding some studies conducted, especially in metropolitan areas and in the case of some selected industries, in the recent past. Industry is prepared to lend a helping hand in this regard.

Monitoring of pollution needs to be standardised through the use of electronic equipment. Pollution testing facilities require a great deal of strengthening Industry itself may set up testing houses on the lines of the Rashtriya Chemical's Laboratory. It will be quite appropriate if these laboratories as well as the national laboratories in various disciplines are associated with the task of evolving pollution standards.

Strengthening of Technical Expertise

The State pollution control boards have also to be strengthened with competent personnel. Multiplicity of the agencies concerned with environmental protection need to be contained through rationalisation. The harassment caused by such factors as inadequate knowledge on the part of the pollution control staff of the intent and purpose of the measures, their educational background being different from the requirements and (sometimes) certain extraneous consideration etc., need to be obviated.

Industry's Representation on Pollution Control Boards

Industry's representation on both the Central and State (prevention and control of pollution) Boards too, is a matter that needs a fresh look. While industries owned by the government are provided specific representation (two members) on both the Central and State board, no such provision is there in their statutes for the representation of the private corporate sector.

There is, of course, a provision for nominating three non-officials on the Boards to represent interests of agriculture and fishery or industry or trade or any other interest which in the opinion of the government (Central Government in the case of the Central Board and State Government in the case of the State Board) ought to be represented. But in practice, most of the State Boards do not have any representation from the private corporate sector.

The private corporate sector has to its credit considerable work in environmental protection. It has also considerable expertise which can be utilised to tackle the problem. It will be quite judicious to make use of this experience and expertise in drawing up practicable measures, especially from the technical viewpoint.

It is equally imperative that the private sector be closely associated in the formulation of environmental protection norms and standards affecting the performance and operations of the particular industries.

Uniform Applicability

The environmental protection measure, it may further be emphasised, ought to be applicable uniformly for all units in a particular line of manufacturer at any particular place whether belonging to the private corporate sector, the state sector, the medium-scale sector, the cooperative sector or the decentralised sector. It is the cumulative impact of the operations of all industrial units that matters and has to be contained and not of the units belonging to a particular sector. Unless all types of units pull together, the requisite results cannot be attained. Besides, as stressed earlier, supportive governmental action is essential. If the local bodies fail to provide back-up infrastructure, as happens not infrequently, all efforts on the part of industrial management can well be negated.

Small Scale Industries

To facilitate matters, especially in the case of small-scale industries, for the development of which industrial estates are being set up in all the states/union territories, and contain pollution control costs, common effluent drainage lines and common treatment plants can be set up. The capital as the maintenance costs can be shared by industrial units if the Industrial Development Corporations provide and operate such drainage lines and plants.

Large Scale Sector Phasing

As regards large units, establishment of primary treatment facilities such as equalisation, neutralisation and clarifier and sludge drying facilities may be taken up in the first instance, going in subsequently for secondary treatment such as aeration. The whole process, of course, may be time-bound-spread over two or three years.

Hazardous Plants and their Neighbourhood

In the wake of the unprecedented Bhopal tragedy, two other issues have come to the forefront. The first is of the neighbourhood of hazardous projects. Care needs to be taken that the highly hazardous plants are located well away from human habitations. These recent guidelines on the sitting of such units. Then, as there is always a tendency for human habitations to grow around industrial establishments, this tendency has to be obviated. The local bodies have been very lax in this regard. Very often, as in the capital as well as in Bombay and Calcutta, large residential concentrations have been even allowed to come up, authorisedly, around the industrial belts which contain several hazardous units.

Since human habitations around industrial units, even if they be highly hazardous cannot always be obviated, the industrial managements should do well by taking measures to educate the habitats about the hazards of their being too close to such industrial projects so that in emergencies, the habitats can take the necessary precautionary steps. The local bodies, too have to shoulder this responsibility in association with industry.

Insurance Issue

The other issue is of ameliorating the distress caused to those affected by mishaps. Indian seeking increased presence in the high technology areas, the hazards are greater because of sophisticated chemicals to be used throughout. It may be worthwhile to consider the setting up of a fund providing relief to the affected neighbourhood to which not only industry but also the authorised habitates may make appropriate contributions the latter in the purchase prices of their habitation itself. Group insurance may be organised through this fund.

Transportation user-end Hazards

Environmental protection, of course, is not to be attempted only at the plan level. Hazards are there in transportation as well as at the user-ends. For instance, apart from the dangers faced by the formulators, their employees and their neighbouring communities, there is a great risk of improper use of pesticides by farmers and their workmen, many of whom are ignorant of the dangers involved and the precautions to be taken in handling the products such as pesticides. Appropriate steps need to be taken at the transportation and user, end levels as well.

Vadodara Experiment

Mention may be made here of the Vadodara experiment in Gujarat, in which, following a study commissioned by the

State Government on the state's ecology, emphasis is placed on ecology and environmental protection as a whole. The cost incurred in studying the problem was shared by all the interests concerned. There was interaction between industry and government at every stage. The general public, particularly professionals and academicans, were involved in the project. Such attempts need to be replicated, as they imply coordinated approach and not a compartmentalised view.

Recycling of Waste Products

Environmental protection may not always prove a cost escalation phenomenon from the long-term point of view. Quite often, the measures towards environmental protection may yield good result and even add to the revenues of an industrial undertaking, especially when valuable materials are retrieved from wastes. These retrieved materials can well be recycled into use. In fact, recovery of various materials from wastes requires to be undertaken on a continuing basis in view of the accentuating shortages of some of these materials and the problem getting even further aggravated in view of foreign exchange stringency.

Quality Production

Industry should do well to devote more attention to the quality of its products. In the competitive market that is being developed in the country, this aspect of production naturally will gain increased recognition. In the road transportation sector, for example, improved fuel efficiency can go a long way in controlling atmospheric pollution. The same is the case with power equipment and irrigation pumps.

Smokeless solid domestic fuel, too, can be developed. A beginning in this regard was made worth the setting up of the Neyveli Lignite Corporation. Experiments in regard to coal have also been going on for many years, but the progress attained leaves much to be desired. The manufacture and popularisation of high efficiency stoves are another such instances.

Environmental protection is not anti-growth. As industrial units move into this programme in big way new manufacturing activities will develop. New demand for waste-reduction equipment is generated and new business opportunities to significant extent.

The programme ought to be drawn up realistic in consultation with the industrial interests concerned, so that they can be held responsible for implementing effectively. There should be national industry-wise norms standards laid down, taking into full cognizance their feasibility, providing some way for State Governments to modify them location-wise and in the light of intensity of pollution. The norms need to be industry and location-specific. All units in a particular industry, irrespective of the sector to which they belong, ought to be treated in a uniform manner. The production of pollution control equipment indigenously should be encouraged energetically and the necessary effort should be put in, where imports are essential, they should be allowed under the OGL and duty-free. Even indigenous production ought to be exempted from fiscal levies-excise duties, sales tax, octroi duties, etc.

Proliferation of agencies dealing with environmental protection ought to be avoided and the concerned agencies should be strengthened with appropriate technical expertise.

The corporate sector's expertise and experience requires to be utilised through grant of specific representation to it on the Central as well as the State (Prevention and Control of Pollution) Boards.

The environmental protection programme has to be an integrated one with the governmental agencies providing adequate support, particularly in respect of infrastructure.

The workforce has also a great role to play in ensuring environmental protection and industrial safety, though the primary responsibility is that of the management. The workforce should, therefore, be educated and trained for discharging its role, particularly in preventive maintenance and proper operations.

Habitations around hazardous units ought to be obviated. The neighbourhood should be imported adequate knowledge of the hazards involved, it ought to be educated to comb the hazards in case of emergencies.

Environmental protection has to be ensured not only at the production level but also at the transportation, marketing and user end levels. It ought to be an integrated and systematic programme.

Retrieval of valuable materials from industrial wastes and effluents and recycling them into the production processes can yield wholesome results. Quality production, too can go on long way in containing pollution.

An ambitious environmental programme can provide stimulus to growth. New manufacturing activities will develop. New demands for waste-reduction equipment will be generated. New business opportunities will arise.

Major Issues

1. Constraints on the implementation of the industrial safety and environmental protection programme by industry so as to rationalise the environmental protection norms and standards as well as the phasing of the implementation of the programme.
2. How far the programme can be made industry and location specific in the larger interest of stimulating industrial growth, particularly of the backward areas?
3. What can be done to contain the agencies for environmental protection in so far as industry is concerned and how these agencies can be strengthened from the point of view of technical expertise?
4. What specific arrangements need to be made to accord representation to private corporate sector on environmental protection agencies?
5. Besides the implementation of the programme, what role should these agencies play in the dissemination of knowledge about environmental protection, especially among the small industrial units?

6. What steps should be taken by industry for sharing of the experience in industrial safety and environmental protection with small industrial units.
7. Are the fiscal incentives which are being provided for environmental protection sufficients? How far do they need to be modified and in what directions?
8. What should be done to stimulate R & D endeavours of the manufacture of pollution control equipments and also to encourage their production?
9. What steps should be taken to safeguard the interests of the neighbourhoods of hazardous industrial units and what sort of insurance arrangement in cases of mishap can be made?
10. What sort of literature should be prepared by industrial experts to disseminate knowledge about environmental protection for general information of public and specific education of industrial workforce?
11. What supportive role the governmental agencies, especially local bodies, have to play in environmental protection by industry and if this supportive role is found wanting, how far industrial units can be penalised for ineffective implementation of the environmental protection programme?
12. What role the industry can be expected to play in containing environmental hazards at the transportation, marketing and consumer-end levels?
13. How far industry can contribute in pollution monitoring and strengthening of pollution testing facilities?
14. What role the apex industry and labour organisations are expected to play in helping their constituents in containing environmental protection?

ANNEXURE-1

ICC Environmental Guidelines For World Industry

First published in 1974, these guidelines were revised in 1981. A second revision was undertaken last year. Reproduced below is the draft of this second revision.

The ICC has prepared the following revised Environmental Guidelines bearing in mind the WICEM (World Industry Conference on Environmental Management) Conference declaration and presents them to governments and companies for implementation and the public for information. Industry's community of nations in regards to the protection and progressive improvement of the human environment is strong as ever. Realising these objectives requires concerted effort by all parts of society and also calls for realism in policies, in the choice of priorities and of time scale, and in the assessment of economic and other costs in reaching specific environmental objectives.

Since the publication of the first edition of its Guidelines in 1974, industries have a accomplished important improvements in its operations to abate and prevent adverse environmental effects. Nevertheless, industry realised that its operations continue to have a substantial, though by no means exclusive, influence on the quality of environment. At the same time, it should be recognised that industry makes a major contribution to the quality of life by providing a regular flow of goods and services.

These Guidelines call for a high standard of commitment to environmental requirements on the part of industry but also make it clear that a constructive contribution from industry depends upon appropriate regulations and governmental policies and indeed upon a parallel contribution from all sectors of the community. For example, the development of alternative sources of energy and the inherent consequences for the environment is one of the issues which will call increasingly for consultation between government and industry. The following general principles are framed.

1. All human activities, both industrial and non-industrial, affect the environment, may lead to pollution and may have an impact on human health and well-being. The universal quiet for higher living standards, together with the rapid growth and urbanisation of world population and the consequent greater dependence on modern industry, requires the continuing attention be given to the consequences on man's activities on the environment.
2. The maintenance and improvement of environmental quality, to the extent consistent with the fulfilment of other human needs, is both essential and feasible; while it entails challenges and often burdens on all elements of society as a whole.
3. In devising environmental protection measures, full account should be taken of the following important factors:
   (a) The importance of human health;
   (b) The need to maintain the balance of ecological cycles, where land, water and air are concerned;
   (c) The need to develop alternatives to non-renewable resources;
   (d) The cumulative effect on the environment of harmful wastes and other nuisances produced by the various industrial operations and their products;
   (e) The existence and impact of transboundary pollution;
   (f) The land of diminishing returns, i.e., inevitably a point will be reached where the benefit derived will be less than a given resource expended; and
   (g) The need to minimise risks to the environmental arising out of industrial activities, bearing in mind that man is always exposed to some degree of risk whether the activity in which he is

involved in industrial, leisure, domestic, or any other.

4. In the field of environmental protection, industry has both economic and social responsibilities.
5. While social and environmental considerations are of great importance, economic constrains will always apply and the aim must be to arrive at a proper balance. Indeed, industry will conduct its operations in an environmentally sound manner, while fulfilling the imperative of economic growth.
6. Improvements in the protection of the environments will best be achieved by applying sound environmental management practices and more technology, improved and tested. Indeed, to a large extent, it is through technology and the use of natural regenerative forces that environmental protection will be achieved.
7. In order to attain worth while results within reasonable time and at acceptable cost, priority should be given to measures for the protection or improvement of the environment that have the highest ratio of community benefit to cost. All environmental policies and regulations require consideration of such cost/benefit analysis, in accordance with the recommendations issued by the ICC in 1980 (publication No. 361, "Cost Benefit Analysis of Environmental Protection Measures").
8. In environmental context, as in others, industry, being an integral part of the community, should-seek always to operate responsibly, taking all reasonable precaution so that its operations and products are consistent with consideration of human health, property, amenities and environment. In the absence of legislation, industry should take independent and responsible action based on its know-how and technology to achieve environmental improvement within reasonable economic bounds.

9. Industry has its particular environmental responsibilities in terms of such factors as plant location and design, process selection and product design, environmental pollution, harmful radiation, vibration and noise controls, occupational health and safety aspects and long-range planning.

10. In the planning and management of its operations, industry should, in addition to the usual elements, take into account the vulnerability of the human environment and of natural ecological systems, as well as the challenge created by the finite character of the earth's non-renewable resources. Industry should, therefore, regularly review its production processes, its procedure for the handling of its materials, its products and its wastes in order to minimise the possibilities of pollution to apply the most appropriate pollution prevention and/or abatement techniques; to conserve wherever possible non-renewable resources, including energy, and to develop alternatives to maximise the recycling or re-use of its wastes; and to develop and utilise cleaner technologies. Any measures envisaged must, of course, also be technically feasible, economically acceptable and consistent with national and/of regional requirements.

11. In the design of its installations and in the operation of its processes, industry should also pay full attention to the protection of the working environment.

12. In the design and development of new products, Industry should assess potential adverse environmental effects of these products when used by a consumers directly or converted or incorporated by other industries into different products. Responsible efforts should be made to minimise any such effect.

13. Industry should provide advice to its customers on conservation, re-use and re-cycling, handling, transporting, using and disposing of its products.

This should encourage the consumer to exercise his responsibilities once he takes possession of the product, in full cognizance of its potential impact on the environment and human health.

14. Each company should promote among its employees an individual sense of environmental responsibilities and should educate and encourage employee at all levels to be alert to potential sources of pollution and to sound resource conservation measures within their operations.

15. Industry has a responsibility through its association channels and at company level to support research into the prevention of adverse environmental effects and into the causes and effects of discharges on ecological systems and on public health, both nationally and internationally.

16. Industry should develop, through its association channels and among individual companies, the exchange of technical information on pollution abatement and conservation methods, subject to commercial confidentiality, proprietary right and patent protection, where applicable.

17. The wide range and complexity of problem raised by environmental protection measure calls for the closest possible contract and consultation between industry and government locally, nationally and internationally, in the search for the mist appropriate solutions. This consultation should include review of the legislative framework for achieving the purpose.

18. Industry supports legislation to achieve environmental improvement which takes into factors local requirements, and which is based on a sound appreciation of technological possibilities.

19. Industry has a responsibility to provide public authorities with available relevant information about

emissions, effluents wastes and other environmental nuisances, including potential adverse health and environmental impacts.

20. Industry should provide public authorities with relevant scientific information and cooperate with them to establish environmental standards and objectives on a sound scientific basis, taking into account confidentiality considerations.

21. Environmental legislation/regulation should be so devise as not to distort international trade relationships and not to create non-tariff barriers. Provisions should be made within existing mechanisms for international consultations to deal with any environmental measures.

22. Any regulatory approach should be based on the establishment of a clearly defined framework for legislative environmental objectives which permits technological flexibility in reaching the desired goal, rather than on the specification of technologies and the composition of materials to be used, i.e., performance standards rather than design limitations.

23. Industry should be given an adequate notice of, and an opportunity to participate in, intended changes in environmental policies, objectives or regulations which might affect its operations. Where changes are adopted, industry must be given a reasonable period of time to adapt them to design limitations.

24. When sitting and designing its installations, industry should be prepared to provide information on steps which it is taking to protect the local environmental and meet safety requirements. In any public debate on issue such as industry should be given a fair opportunity to state its case. The aim must be to reach solutions mutually acceptable to industry, the authority and the immunity.

25. Industry and public authorities should jointly work out contingency plans to deal with pollution

emergencies and accidents, industry should inform the relevant authorities about the known and significant hazards of its operations, so as to enable them to act quickly and properly.

26. Industry should take energetic steps to ensure that its very substantial achievements in matters of the environment are brought continuously to the notice of the public in general.

27. Industry should take into account public opinion and opinions of scientist and other concerned organisations when developing and implementing environmental protection programmes in each particular case and take the lead in raising the level of awareness and understanding of these programmes.

28. Industry should be prepared to supply information about possible environmental consequences of its operations, products and new developments.

29. Industry should provide input to balance and inform public discussion of environmental problems and should support efforts to place in proper perspective the comparative significance of industrial and non-industrial sources of pollution.

6

ENVIRONMENTAL IMPACT ASSESSMENT

Environmental Impact Assessment is the analysis of any possible alterations of environmental conditions, adverse or beneficial, caused introduced by the action or set of actions under consideration. It is anticipatory in nature. Ideally, in environmental impact assessment should be prepared along with the project report. But in many cases this is not done. In this paper a detailed presentation is made on what an EIA should attempt. The starting point of an EIA is a preliminary assessment.

Preliminary Assessment

A preliminary assessment will attempt to answer the following questions:

Are there potential physical or health hazards associated with location of the proposed facility?

Will there be significant disturbance on the existing communities?

Are there potential conflicts with other economic interests?

Will the project require major additions to existing physical infrastructure including transport, water and housing?

What would be the increased demand upon the existing provision of social services?

What will be the future growth pattern of that region in the next 20 years?

A preliminary assessment will indicate the possible impact are as on which detailed data has to be collected-for presenting the results of the preliminary assessment, an impact matrix is prepared. Table 6.0 gives the results of a hypothetical preliminary assessment. A preliminary assessment identifies the areas to be considered in the environmental impact assessment process. After the preliminary assessment a detailed plan is made for preparing an environmental impact assessment covering all major aspects.

Environmental Impact Assessment

The following is a model format for environmental assessment, suggested by UNEP, namely

Step 1: A description of the project in detail and review the data base.

Step 2: Listing of all the resources used directly or affected indirectly by the project and residues created.

Step 3: Listing and quantitative assessment of the adverse effects of the project in terms of the resources which may be exhausted, depleted or deteriorated.

Step 4: Listing and quantitative assessment of the resources such as land, water and public health enhanced as a result of the project.

Step 5: Identification and quantification of project components required for conserving, preserving or restoring the resources affected by the project. Monitoring systems that are to be implemented for the continuous monitoring of the impact related parameters.

Step 6: Forecast short-term and long-term environmental changes arid identification of modifications.

Step-1 will consist of project description of all constituent activities from land acquisition to development of operational aspects including alternative.

Step-2 also gives the site description, present socio-economic and environmental status and site requirements for project implementation.

Step-3 involves enumeration of impacts such as physical, chemical, health, ecological and socio-economic, both adverse and beneficial, evaluation of impacts in quantitative terms or qualitative terms and aggregation of impacts on environmental components such as land, water, air, forest, flora, fauna, settlements, health etc.

Given, this background information, the major impact areas that are to be covered are highlighted below.

Background Information

The first step in carrying out a detailed impact assessment is to generate base line data on the following parameters:

1. Land use patterns
2. Socio-economic parameters
3. Soil characteristic
4. Agricultural production patterns
5. Watershed description
6. Meteorological and Wind characteristics
7. Topography in the vicinity
8. Sensitive areas in the vicinity
9. Human settlements
10. Vegetation characteristics
11. Ground water regions and acquifiers
12. Surface flow of streams and seasonality

13. Fauna and Flora in the vicinity
14. Wastelands and possible land fill areas.

It is not necessary to generate all these by the concerned agency. All these information are available with various agencies and can be collected without difficulty.

The possible environmental impacts is an area that needs detailed analysis and forms the crux of the environmental impact assessment. Either a multidisciplinary group can be appointed or a multidisciplinary consulting team can be approached for undertaking this part of work. A checklist method is normally used for identifying the possible impacts in the construe and operating phase.

In the case of industries causing air pollution, meteorological characteristics are critical variable. A detailed description of the weather which includes diornal and seasonal patterns (wind direction and wind speed), atmospheric stability inversion heights and their frequency of occurrence, is necessary. One of the first things be prepared is a wind rose for describing the log wind speeds and directions. The detailed procedure for preparing wind roses is given in Fig. 6.3. Then data on atmospheric surface stability collected. Surface wind speed, solar isolation and cloudiness will determine the extent of primary pollutant impact. The pasquill chart is used for determining atmospheric stability values for a location.

Another base line data that is to be collected is existing air quality. An assessment of existing air quality is necessary for prediction of the future impact of projects. Two kinds of data may be needed in this regard namely-(i) regional air quality, and (ii) subregional air quality. If no data are available monitoring may be needed for estimating the impact on air quality due to changes in the meteorology and source strength.

Once the base line data are obtained detailed impact assessments can be carried out.

Table 6.0 : Illustrative Ad Hoc Approach to Environmental Impact Versus Environmental Area

Environmental Impact *Environmental Area*	*No effect*	*Positive effect*	*Negative effect*	*Bene-ficial*	*Adverse*	*Proble-matic*	*Short term*	*Long term*	*Rever-sible*	*Inrever-sible*
Wildlife			x			x	x			
Endangered Species	x									
Natural Vegetation			x			x			x	
Exotic Vegetation	x									
Grading			x			x			x	
Soil Characteristics	x									
Natural Drainage	x									
Groundwater		x		x						
Noise			x				x			
Surface Paving						x				
Recreation	x									
Air Quality			x		x			x		x
Visual Disruption	x									
Open Space			x		x			x		x
Health and Safety	x									
Economic Values		x		x				x		
Public Facilities (includes schools)						x	x	x		
Public Services	x									
Conformity to Regional Plans		x		x				x		

Fig. 6.1. Musa velutina wendl. & Drude (Wild bana) Arunachal Pradesh (note the erect natre the inflorescence).

Air Quality Impact Analysis

A detailed air quality impact analysis will contain the following:

1. Description of the project area
2. Meteorology

   ..wind rose

..atmospheric surface stability

..inversion height

3. Existing air quality

..Local

..Regional

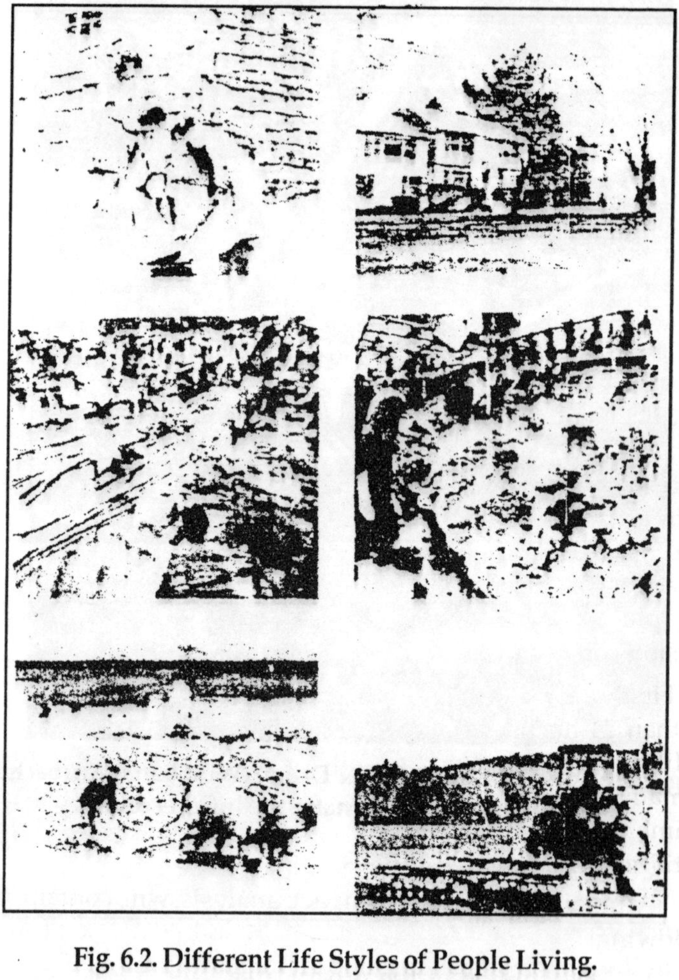

Fig. 6.2. Different Life Styles of People Living.

4. Detailed aspects of the project

..Description of the vicinity

..Traffic study

5. Impact scenarios

..Emission scenarios

..Meteorological scenarios

..Air chemistry scenarios

..Primary and secondary impact scenarios

6. Air quality modeling

Generation of impact scenario and air quality modelling are the most important steps in the air quality impact analysis.

The dispersion process is dependent on mixing and dilution of waste gases in the atmosphere. For estimating dispersion a parameter known as effective stack height is estimated. It is the actual stack height plus the plume rise. The dispersion is assumed to start at the effective stack height. The calculation of the ground level concentration is based on a two step dispersion process. The pollutant gases from a stack source rise up as a result of thermal conditions and then dispersed in accordance with a Gaussian (bell shaped) distribution. Wind speed, wind shear and eddy currents influence the interaction between a plume and atmosphere. As a practical step, plume rise is assumed to be a function of only the emission conditions at the time of plume release.

Firstly, short-term ground level concentration for periods less than 24 hrs are estimated, assuming that meteorological conditions will remain constant for the prediction period. Normally exist temperature of the stack gas is greater than the ambient temperature, and because of this, the plume will rise till it is cooled by the surrounding air. This is known as thermal plume rise. Gases and particles of less than 20 microns will disperse at the effective stack height and then spread out in both the planes (horizontal as well as vertical planes). The basic dispersion equation used for calculating the ground level concentration directly downwind from the emission point is as follows:

$$X = \frac{Q}{\prod u_s\ \sigma y\ \sigma z}\ EXP\left[\frac{h^2}{2\sigma z}\right]$$

where X = ground level conce. (gm/m^3)

Q = emission rate at source (gm/sec)

h = effective stack height (m)

u_s = mean wind speed at (m/sec) stack height

σy, σz = Horizontal and vertical plume standard deviations (m)

The maximum ground level concentration is given by the following equation:

$$X_{\max} = \frac{2Q}{e\prod u_x\ h^2}\ \frac{\sigma z}{\sigma y}$$

For calculating h. The plume rise (Δh) is calculated for various conditions depending on the stability of the atmosphere.

For stable atmospheric conditions the following equation is used:

$$\Delta h = 2.9\left[\frac{2.45V_s\ D^2\ (T_s - T_a \mid T_s)}{u_a\ (9.8/Ta)\left[\frac{\Delta T}{\Delta Z} + 0.98\right]}\right]$$

where

Δ h = plume rise (m)

D = stack diameter (m)

V_s = stack gas exit velocity (m/s)

U_s = mean wind speed at stack height hx (m/s)

T = exit temperature of stack gas (°K)

T = ambient temperature (°K)

For neutral and unstable atmospheric conditions another equation is used

$$\Delta h = \frac{7.4h_s^{2/3}}{u_s}[2.45V_s\ D^2\ (T_s - T_a)/T_s]$$

These equations are used when exit temperature stack gas are 50°K greater than ambient temperature and large volumes of gases are belong discharged.

Computer simulation is based as a technique for calculating ground level concentrations at various emission rate, velocity or temperature. Computer programmes are available for this purpose which include:

(1) receptor site and stack coordinates

(2) stack coordinates

(3) desired meteorological data

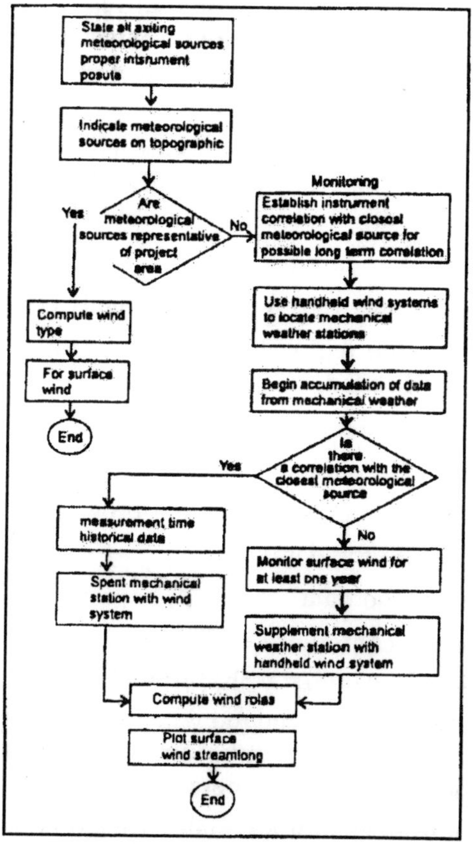

Fig. 6.3. Flowchart for Wind Rose Preparation.

Large number of simulation exercise are carried out by varying pollutant emissions, stack height, stack diameter, exit gas velocity, exit gas temperature, exit gas molecular height, wind speed, wind direction, atmospheric stability and wind averaging time. After estimating the ground level concentrations (average and max) the long-term average ground level concentrations are estimated. The long-term average ground level concentrations are estimated using actual meteorological data to predict monthly, seasonal and annual average ambient concentrations. The weather data are derived based on thousands of observations of wind speed, wind direction, and atmospheric stability taken over the desired averaging interval. Using computer simulations hundred of computations are carried out to generate the long-term average ambient concentration at various locations.

In certain cases the data may not be readily available. In such cases a preliminary air quality survey may be needed. The steps for generating the base line data may be the following:

1. Identify the data needs
2. Establish the need for the air quality survey
3. Identify the region to be surveyed. A mobile unit may be needed.
4. Determine the pollutants to be sampled
5. Identify the sampling instrumentation
6. Determine a proper sampling programme
7. To acquire the data
8. Analyze the data
9. Conduct computer simulations for short-term and long-term average concentration
10. To make recommendations concerning impact of the proposed project, optimum stack height and air quality monitoring programme.

The major exercises for air quality should be with regard to sulphur dioxide, particulates and trace in the ash, especially respirable particles.

Water Quality Impact

Water quality impact is very essential in terms of the availability, quality, variability etc. Quality of water may be affected in a number of ways namely

(i) thermal pollution of the receiving waters;

(ii) return of dissolved oxygen deficient waters;

(iii) disposal of suspended solids, dissolved solids, toxic materials;

(iv) run off from solid wastes

(v) seepage of polluted water into ground water.

The waste water survey should identify the following:

(i) locate outfalls;

(ii) characterize outfall effluents;

(iii) characterize effluents characteristics;

(iv) drive impact of effluents.

Dissolved oxygen concentration need attention. Attention should be given to estimate the impact due to change in ionic strength, dissociation, conductivity and solubility of chemicals in the zone affected. The environmental indicators that are to be monitored are:

(i) temperature

(ii) dissolved oxygen

(iii) total dissolved solids

(iv) nitrogen

(v) heavy metals

(vi) thermal stratification, and

(vii) toxic substances.

Impact of water quality on the sources and receiving waters have to be performed separately. This has to be done for various seasons separately. For predicting oxygen concentration either mixing zone models or dissolved oxygen models are used. For prediction the temperature impact, temperature models are used. For predicting the effect of run off there are empirical models available.

The starting point of water quality impact assessment is the preparation of the microwatershed map. Simultaneously stream flow rates, aquifer reserves and effluent characteristics are estimated. The impact analysis is carried out using the appropriate models. A topographical map, hydrological map, and a soil characteristics map are also to be prepared. These maps are available with the concerned government agencies.

The analysis of potential impacts have to be derived since the percentage of water used as a proportion of ultimate utilizable potential has been going up. Some of equations used for predicting water quality impact are given below:

$$C_r = \frac{C_w Q_w + C_u Q_u}{Q_r}$$

C_r = Concentration of the pollutant in the river down-stream (mg/1)

C_w = Concentration in waste discharge (mg/1)

C_u = Concentration in upstream river (mg/1)

Q_u = Upstream river flow (m^3/Sec)

Q_w = Waste discharge rate (m^3/Sec)

$Q_r = Q_u + Q_w$ = river flow at downstream station (m^3/sec)

For predicting the minimum dissolved oxygen the following equation $_2$ can be used:

$$DO_{\min} = DO_s - \frac{K_1 L}{K_2} e^{k_1} tc$$

DO_{min} = Minimum dissolved oxygen

DO_s = Dissolved oxygen concentration if the water were saturated

K_1 = deoxygenation rate constant

K_2 = rate of reaeration

t_c = Time in days at which dissolved oxygen reaches its minimum value.

L = Initial ultimate BOD concentration

For predicting the temperature of receiving water the following equation can be used. This is applicable only in the case of a well mixed body of water.

$$T_s = T_n + \frac{H_p}{AK}$$

Ts = Water temperature with heated discharge;

Tn = Natural water temperature before heated discharge;

Hp = Plant heat rejection rate (Watt)

A = Surface area of the water body (m^2)

K =Surface heat exchange coefficient (W/m^2)°C.

Another parameter that requires attention under the Indian conditions are the likely flood water flows and the sufficiency of drainage. The possibility of floods and its likely impact should be incorporated. If areas are prone to frequent floods sufficient safeguards are necessary and these safeguards should be identified at the EIA stage itself.

Any industry is likely to create a township as well as allied economic activities. These additional human settlements may also generated sewage and other solid wastes. The possible impact of this on the receiving water may also have to be carried, even if treatment is envisaged in the project.

It is necessary to assess the possible impact on the biological environment namely human settlements, vegetation, wildlife and aquatic life. This should be based on data that can be obtained from Botanical and Zoological Survey of India. Aquatic Life may be easily affected by waste water discharge. A survey of aquatic resources and their utilization will have to be studied. In the survey of aquatic resources, spawning grounds of fishes may have to be identified and also the spawning periods. Thermal pollution has severe impact on aquatic life and the discharge of cooling tower water has to be done such that even under adverse ambient conditions the temperature of the outfall water is not higher than 5°C. Detailed computer simulatic exercises may be necessary while designing the cooling tower. Enough allowance may be made for increase in ambient temperature during summer, reduction in the stream flow in summer and the increased water requirements in summer.

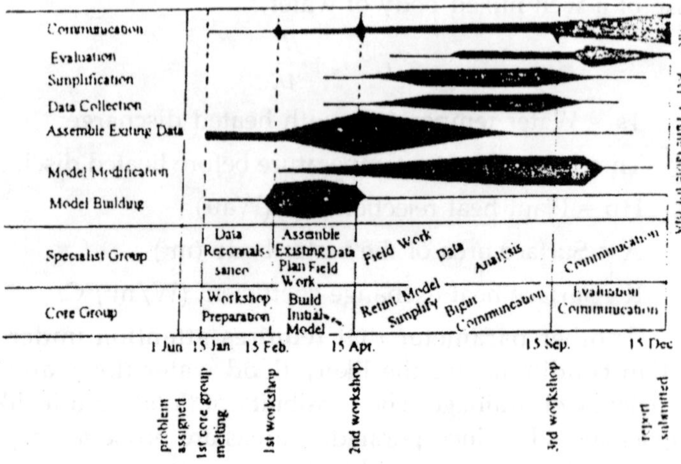

Fig. 6.4. Any industry is likely to create a township as well as allied economic activities. These additional human settlements may also generate sewage and other solid wastes. The possible impact of this on the receiving water may also have to be carried, even if treatment is envisaged in the project.

Table 6.1. Project Actions

A. Modification of Regime

(a) Exotic flora or fauna introduction
(b) Biological controls
(c) Modification of habitat hydrology
(d) Alteration of ground cover
(e) Alteration of groundwater
(f) Alteration of drainage
(g) River control and flow codification
(h) Canalization
(i) Irrigation
(j) Weather modification
(k) Burning

(l) Surface or paving
(m) Noise and vibration

(B) Land Transformation and Construction

(a) Urbanization
(b) Industries and buildings
(c) Airports
(d) Highways and bridges
(e) Roads and trails
(f) Railroad
(g) Cables and lifts
(h) Transmission lines, pipelines and corridors
(i) Barriers including fencing
(j) Channel revetments
(k) Channel dredging and
(l) Straightening
(m) Canals
(n) Dams and impoundments
(o) Piers, seawalls, marines and sea terminals
(p) Offshore structures
(q) Recreational structures
(r) Blasting and drilling Cut and fill
(s) Funnels and underground structures

C. Resource Extraction

(a) Blasting and drilling
(b) Surface excavation
(c) Subsurface excavation and retorting
(d) Well drilling and fluid removal
(e) Dredging
(f) Clear cutting and other lumbering
(g) Commercial fishing and hunting

D. Processing

(a) Farming
(b) Ranching and grazing
(c) Feed lots
(d) Dairying
(e) Energy generation
(f) Mineral processing
(g) Metallurgical industry
(h) Chemical industry
(i) Textile industry
(j) Automobile and aircraft
(k) Oil refining
(l) Food
(m) Lumbering
(n) Pulp and papery
(o) Product storage

E. Land Alteration

(a) Erosion control and terracing
(b) Mine Sealing and waste control
(c) Strip mining rehabilitation
(d) Landscaping
(e) Harbor dredging
(f) Marsh till and drainage

F. Resource Renewal

(a) Reforestation
(b) Wildlife stocking and management
(c) Ground water recharge
(d) Fertilization application
(e) Waste recycling

G. Changer in Traffic

(a) Railway
(b) Automobile
(c) Trucking
(d) Shipping
(e) Aircraft
(f) River and canal traffic
(g) Pleasure boating
(h) Trails ponds
(i) Cables and hits
(j) Communication
(k) Pipeline

H. Waste Replacement and Treatment

(a) Ocean dumping
(b) Landfill
(c) Emplacement of tailings, spoil, and overburden
(d) Underground storage
(e) Junk disposal
(f) Oil well flooding
(g) Deep well emplacement
(h) Cooling water discharge
(i) Municipal waste discharge including spray irrigation
(j) Liquid effluent discharge
(k) Stabilization and oxidation
(l) Septic tanks, commercial and domestic
(m) Stack and exhaust emission
(n) Spent lubricants

I. Chemical Treatment

(a) Fertilization
(b) Chemical de-icing of highways, etc.

(c) Chemical stabilization of soil

(d) Weed control

(e) Insect control (pesticides)

J. Accidents

(a) Explosions

(b) Spills and leaks

(c) Operational failure

Table 6.2. Environmental Characteristics and Conditions

A. Physical and Chemical Characteristics

1. Earth

(a) Mineral Resources

(b) Construction material

(c) Soils

(d) Land form

(e) Force fields and background radiation

(f) Unique physical features

2. Water

(a) Surface

(b) Ocean

(c) Underground

(d) Quality

(e) Temperature

(f) Recharge

(g) Snow, ice, and permafrost

3. Atmosphere

(a) Quality (gases, particulates)

(b) Climate (micro, macro)

(c) Temperature

4. Processes

(a) Floods

(b) Erosion

(c) Deposition (Sedimentation, precipitation)

(d) Solution

(e) Sorption (ion exchange, complexing)

(f) Compaction and settling

(g) Stability (slides, slumps)

(h) Stress-strain (earthquake)

(i) Air movements

B. Biological Conditions

1. Flora

(a) Trees

(b) Shrubs

(c) Grass

(d) Crops

(e) Microflora

(f) Aquatic plants

(g) Endangered species

(h) Barriers

(i) Corridors

2. Fauna

(a) Birds

(b) Land animals including reptiles

(c) Fish and shellfish

(d) Benthic organisms

(e) Insects

(f) Microfauna

(g) Endangered species

(h) Barriers

(i) Corridors

C. Cultural Factors

1. Land use

(a) Wilderness and open spaces

(b) Wetlands

(c) Forestry

(d) Grazing

(e) Agriculture

(f) Residential

(g) Commercial

(h) Industrial

(i) Mining and quarrying

2. Recreation

(a) Hunting

(b) Fishing

(c) Boating

(d) Swimming

(e) Camping and hiking

(f) Picnicking

(g) Resorts

3. Aesthetics and human interest

(a) Scenic views and vistas

(b) Wilderness qualities

(c) Open space qualities

(d) Landscape design

(e) Unique physical features

(f) Parks and reserves

(g) Monuments

(h) Rare and unique species of ecosystems

(i) Historical or archaeological sites and object

(j) Presence of misfits

4. *Cultural slants*

(a) Cultural patterns (life style)

(b) Health and safety

(c) Employment

(d) Population density

5. *Man-made facilities and activities*

(a) Structures

(b) Transportation network (movement, access)

(c) Utility networks

(d) Waste disposal

(e) Barriers

(f) Corridors

D. Ecological Relationship Such As:

(a) Salinization of water resources

(b) Eutrophication

(c) Disease-insect vectors

(d) Salinization of surficial material

(e) Brush encroachment

Environmental impact statements are meant to be action forcing mechanisms to ensure compliance of various environmental regulations. EIS has to be a useful document where the following criteria may have to be observed:

(i) The EIS should be unbiased and objective;

(ii) EIS should be carried out with the help of a multidisciplinary team;

(iii) The EIS preparation should be started along with the preparation of project feasibility report;

(iv) The safeguards identified in the EIS should be incorporated into the project design itself and for this EIS and project design are to be complementary.

The project initiators normally tend to minimize the cost of EIS preparation, probably sacrificing the quality. The deleterious environmental impact of projects will be greater if environmental factors are only considered in a EIA prepared after the preparation of the project. A tentative time schedule and work plan for the preparation of an Environmental Impact Statement are to be framed according to project at sight.

Air Pollution Control

Air pollution is the presence of any substance or combination of substance that is detrimental to health of man or other forms of life. Air pollution is another example of tragedy of the commons. A common pool resource (social assets) like air is converted into private profit and no one wants to purify or clean it, but everyone is eager to use it. In nature, the presence of natural cycles make it sure that there is a fair balance. Industrialization, large-scale transportation combustion of fuels, and other natural processes introduce substances that are detrimental to humans, animals, plants and material substances. Earlier, pollutions of air was not considered a major problem, but recent evidences indicate that the combined effect of discharge of pollutants from industries, and urbanization, have impact on human health in dust particles are the major pollutants, air pollution was recognised as a problem after the smog incident in London in 1952 in which there were many casualties.

Air pollution and its possible consequences have received recently serious attention. One of the most serious consequences of air pollution is possible climatic changes generated by the increasing concentration carbondioxide in the atmosphere. Carbondioxide produces the so called "green house effect" by retaining long wave radiations. This can possible induce climatic changes and this will have maximum impact on highly populated countries like India, Bangladesh, Pakistan, Indonesia, etc. These countries are tropical and are sensitive to monsoon changes. Here again, the carbondioxide increase is a consequence of two opposing trends upsetting the existing natural balance. Burning fossil fuels such as coal and petroleum and agricultural residues, fuelwood, refuse and animals wastes introduce large quantities of carbondioxide into the atmosphere, whereas the forests which fix carbondioxide in nature is getting severely depleted. The consequences has been a sharp rise in the carbondioxide concentration has been stated that carbondioxide with induce several climatic change in the statements have been contradicted in risk terms. Climatic changes can have severe consequences on the countries like India since the consequences can be highly unpredictable irreversible oxide is not classified as a pollutant but it can induce a large scale changes which can not be corrected if the consequences start manifesting.

Air Pollution in India

Air quality monitoring is done only in selected cities in India. Since the air quality data has to be collected systematically at a regular intervals in specific monitoring stations, available information does not completely depict the status of air quality degradation over a period of time. However, in certain areas the air quality is below the permissible levels. National Environmental Engineering Research Institute. Nagpur has till today brought out air quality in selected cities. The table 6.3 gives the concentration of sulphur dioxide in microgrammes in cubic meter of air at some selected areas in major cities. This shows the following.

- In most of the industrial areas concentration of sulphurdioxide is above the permissible values,
- Even in residential areas the concentration of sulphurdioxide is comparatively high in cities such as Calcutta, Delhi and Bombay,
- Long-terms trends in air quality cannot be analyzed because of lack of data,
- Even in the cases of selected cities, different surveys. have covered different location. Which makes it difficult to make comparative analysis of air quality.

Air Pollutants

Before going into the details of impact of air pollution. Characteristics of some major pollutants are examined. Table 6.3 gives a classification of pollutants.

Table. 6.3. Sulphur Dioxide Levels

(Daily maximum in microgrammes per cubic meter)

	1988	1989	1990	1991
Bombay	444	1084	312	395
(Industrial Area)	(Chembur)	(Chembur)	(Parel)	(Parel)
Bombay	310	286	286	310
(Residential Area)	(Khar colony)	(Khar colony)	(Bandra)	(Bandra)
Calcutta	451	415	396	316
(Industrial Area Cassipore)				
Calcutta	315	206	274	389
(Bhawanipore Residential Area)				
Delhi	674	335	351	325
(Industrial Area: Najafagarh)				
Delhi	267	230	341	245
(Residential Area: Netaji Nagar)				
Hyderabad	280	260	417	230
(Industrial Area: Noula Ali)				
Hyderabad	267	268	254	248
(Residential Area: Tarnaka)				

Sulphur Oxides

Oxides of Sulphur is a major air pollutant since it affects humans, animals and materials. Sulphur dioxide is the major pollutant under this category and it gets converted into sulphur trioxide. These oxides in the presence of moisture produce sulphurous and sulphuric acid. Oxides of sulphur and these acids react with tissue of plant and animal origin. The main problem with oxides of sulphur is that they are transported across large distances and during the transport across large distances and sunlight and other catalysts speed up the conversion of sulphur dioxide into sulphuric acid. The sequence is

$$2SO_2 + O_2 \rightarrow 2SO_3$$
$$SO_3 + H_2O \rightarrow H_2SO_4$$

The major sources of sulphur dioxides are coal combustion, combustion of petroleum product, refuse burning, refinery operations and metallurgical operations.

Nitrogen Oxides

There are a number of oxides of nitrogen generated and because of the similarity of these, they are generally denoted as NO_x. At lower temperatures nitric oxide is produced at higher temperature this gets converted into nitrogen dioxide.

The major source of nitrogen oxides is automobiles, furnaces, boilers, etc. Gas fired furnaces and coal fired furnaces and power station are common source, a combustion of wood as well as refuse also produce nitrogen oxides and ammonia.

Carbon Monoxide

This is an established pollutant and it is mainly generated by motor vehicles.

Carbon monoxide is also produced as result of smoking tobacco.

Mandatory emission inspection and maintenance programmes that involve simple emission inspection tests and engine adjustments may have to be introduced to improve the air quality in urban areas in a short time period.

Exposure lasted 10 min effect on flow resistance was observed 70 min after termination of exposure.

Effect was produced at this concentration in normal subjects with chronic respiratory disease.

Effect occurred in subjects at rest with chronic respiratory disease.

Exposure lasted 10 min: maximal effect on flow resistance was observed 30 min later.

Some estimates of vehicle emission in some major cities are given in Table 6.4.

Table 6.4. Summary of Human Response to Short-term No $_2$ Exposures

Effect	*NO, concentration Mg_2/m_3*	*ppm*	*Time to effect*
Odor threshold	0.23	0.12	Immediate
Threshold for	0.14	0.075	Not reported
Dark adaptation	0.50	0.26	Not reported
Increased airway	1.3-3.8	0.7-2.0	20 min
Resistance	3.0-3.8	1.6-2.0	15 min
	2.8	1.5	45 min
	3.8	2.0	45 min
	5.6	3.0	45 min
	7.5-9.4	4.0-5.0	40 min
	9.4	5.0	15 min
	11.3-75.2	6.0-40.0	5 min
Decreased palmonary diffusing capacity	75-9.4	4.0-5.0	15 min

Acid Rain

A major consequence of air pollution is acid rain. The pollutants emitted come back along with rain, thereby affecting soil, crops, plants, lakes, etc. Acid precipitation is formed in the atmosphere through conversion of nitrogen oxides and sulphur dioxide. Nitrogen oxides and sulphur dioxide are transformed into sulphuric and nitric acids through oxidation. The longer, sulphur dioxide and nitrogen oxides remain in the atmosphere, the greater the likelihood that they will oxidize and become acids. Acid precipitation have a number of potential effects on terrestrial ecosystems, namely:

Acidification of soils, altered nutrient supply, increased mobilization of aluminium and other shifts from acid sensitive to acid tolerant species of microbial and soil fauna population is the soil, altered rates of decomposition of organics and nitrogen fixation;

- foliar damage to crops and forests, interference with productive processes, leaching of nutrients and other substances from leaves, increase or decrease germination depending on sensitivity to acidity;
- indirect effects on agricultural crops include fertilization by nitrates and sulphates, promoting growth;
- potential indirect effect on forests include acidification of forest soils and accompanying alterations in soil chemistry as well as reduced forest productivity and forest, dieback;
- potential effects on aquatic systems include acidification, decreased alkality and mobilization of aluminium and other metals;
- other biological effects on aquatic biota include altered species composition among plankton, vegetation and invertebrates, reduced population of decomposer organism is, decline in productivity of amphibians and fish skeletal deformities and increased mortality of fish;
- increased fish kills may occur during heavy rains.

Table 6.5. Vehicle Emission Inventory for Sulphur Dioxide Lead and Particulate Emissions

Cities	*Type of Vehicle*	*1990-1991* SO_2	*pb*	*Particulate*	*1992-1994* SO_2	*pb*	*Particulate*
Delhi	Gasoline	474	40.1	-	686	73.2	-
	Diesel	8727	-	837	13,906	-	3,071
	Total	9,001	5001	837	13,792	73.2	'3,071
Bombay	Gasoline	512	49.4	-	756	72.5	-
	Diesel	8,331	-	800	12,640	-	3,008
	Total	8,644	49.4	800	13,196	725	3,008
Calcutta	Gasoline	301	36.1	-	379	48.7	-
	Diesel	4,253	-	414	5,786	-	559
	Total	4354	36.1	414	5,965	48.7	559

Acid rain is highly interactive problem and the remedial measures to control it are prohibitively costly. Only of the main arguments against thermal power generation option is the potential of acid rain. It is again, desirable to control the pollution at the source rather than treating the effects of acid rain.

Air Pollution and Forests

Forests can act as sinks for pollutants as well as source of pollutants. Forests fire produces source of quantities of pollutants. Air pollutants and forests interest in three ways:

(i) Under conditions of low dose of contaminants the vegetation and soils of forest ecosystem function as important sources and sinks for air pollutants,

(ii) When exposed to intermediate dose individual tree species or individual members of a given species may be subtly and adversely affected by nutrient stress, and impaired metabolism,

(iii) Exposure to high dose may induce act, morbidity or mortality of specific species.

The fact that forest ecosystems act as a sink for air pollutants at low concentration can be taken advantage for pollution control. There is substantial evidence that trees remove gaseous pollutants. Pollutants like hydrogen fluoride, sulphur dioxide, nitrogen dioxide and ozone which are soluble are readily removed. Nitric oxide and carbon monoxide which are insoluble are not removed by vegetation. Further, pollutants are absorbed most efficiently by plant foliage near the canopy surface. Leaf surfaces accumulate a variety of natural and anthropogenic particles from the atmosphere smaller leaves are generally more efficient particles collector than larger leaves. Smooth leaved species are less efficient than rough leaved species. Conifers are generally more effective particulate sinks than deciduous sinks. Trees act as efficient filters for air borne particles because of their large size, high surface to volume ratio of foliage, petioles, and twigs and frequently hairy or rough leaf, twig or bark surfaces.

Recently in India, large scale mining, thermal power generation and setting up of cement and quarrying units have been taken up of close to forests. This is finally to have adverse affect on forests. Some of the effects of pollutants in medium and high doses are summarized below:

- Reproduction of forest trees may be affected by a variety of pollutants.
- Pollution affects pollen and result in reduced pollen germination.
- Young tree seedlings are affected severally by pollutants and results in stunted growth.
- With foliar symptoms including chlorosis, silvering and other discoloration, may occur from average concentrations of sulphur dioxide in the range of 0.008 ppm to 0.017 ppm (i.e. 21 to 44 microgramme per cubic metre).
- Nitrogen dioxide exposure for more than 10 hrs 1 ppm can cause leaf injury.
- Forest ecosystems may have severe and dramatic impact when subjected to high doses of pollutants. Plants close to smelters and steel plants can be examples for this exposure. Even upto 30 kms away from the source there can be severe effects.

To conclude forests can protect humans from pollutants at low doses, but at high forest sink itself and is destroyed by pollution.

Some of the major issues in air pollution control are given below:

(i) As of now there are no regulations for controlling pollution from mobile sources which is comprehensive.

(ii) Old industries or units have no space or capital to install pollution control devices.

(iii) Under adverse meteorological conditions, there can be serious location specific problems if at the time of sitting not enough attention is given to dispersion of pollutants.

(iv) In areas where ambient levels of pollutants are high, fixing of emission levels should take into consideration the existing ambient air quality.

(v) Air pollution control has to deal with pollution at the source of pollution at the generation point from equipments. For example, without improvements in vehicle engines, vehicular emissions cannot be drastically reduced. There has to be new source performance standards for pollution generation equipments.

(vi) Air quality monitoring needs standardization and institutionalization. For this, the facilities at the Pollution Control Boards need strengthening. Pollution monitoring and regulation agencies should not be the same.

(vii) Pollution control at work places (occupation health) has not received any attention among the policy makers.

(viii) Air pollution from small-scale units seems to be a major problem in a number of areas (for example Mathura Refinery an impact of foundaries on air quality). There seems to be very little efforts to control this problem. It is likely to be a serious problem in the years to come. This has received very little attention and no major programme are currently underway to reduce the magnitude of the problem. Small units have no capital, no interest, no commitment and no expertise to handle the air or for that matter water pollution problems.

(ix) Non-availability of high quality equipment for air quality monitoring is another major problem so also non-availability of soft loans for installing pollution control equipment in industries.

(x) There is very little awareness about the nature of the air pollution problems as well as its consequences, even among experts.

(xi) Though emission regulation, with respect to air has been brought out, the implementation has been weak. Firstly, pollution control boards are not geared to handle the complex issue and air pollution control will need large scale investments on control equipment. Large-scale investments are likely to be made without some institutional changes for financial credit.

(xii) There has been a mushrooming growth of small-scale industries and this segment has not been doing anything to control air pollution. Unless something comprehensive is done to handle the problem, this segment will continue to degrade the environment.

Cities are expanding and our industrial base is developing rapidly. These will inevitably increase the discharge of pollutants to the environment. Air pollution is intricately linked to dispersion of pollutants and meteorological conditions. Unless these conditions are considered at the time of selection of sites. The problem cannot be mitigated by only handling the pollution at the source. Sitting criteria for industrial location has to be made rigorous and no exemption should be permitted. There should be close coordinating of urban planning and industrial location. In certain areas, under adverse meterological conditions there can be problem, and no industries which can cause pollution may be permitted.

Another point which requires attention is new source performance standards so that the problem can be dealt with at the source itself. Unless this is done, the vehicular pollution will not come down drastically. Promoting two stroke petrol engines and their concentration in urban areas are going to be very intricate problem that cannot be very easily resolved because of poor growth of public transport facilities in urban areas.

Air quality monitoring facilities have to be set up in major industrial areas such as Trombay (Bombay) Baroda, Vizag, Sigarauli, Calcutta, Madras, Independent agencies should be set up and these should not be under the pollution control boards.

There seems to be very little awareness among people regarding the consequence of pollution on human health. The only detailed study available on linkage between health and pollution is for Bombay. Such studies need to be undertaken. Impact of airborne fine particles on health and the environment has to be studied in detail. The technical expertise of Pollution Control Board to handle air pollution problems needs, strengthening in terms of training and facilities.

Implementation of air pollution control regulations has to be rigorously done. For this, a detailed dialogue may be initiated with major Industry Associations, senior executives and technical experts. The implementation strategy adopted may have to reviewed with a view to increase its effectiveness.

Finally, in and around industries green belts should be planted and developed. In the vicinity of urban areas and in urban centres there has been sharp reduction in the tree cover. A systematic programme has to be initiated to develop urban forestry and green belts around urban centres to absorb some of the released pollutants.

Water Pollution in Urban India

Indian cities have been growing at an alarming rate over the past few decades. This trend of urbanisation is a natural corollary of the process of economic growth through industrialisation. The cities in India consequently are facing the traditional environmental problems such as lack of sanitation, water supply, chronic shortage of services, polluted air and water, lack of open space and recreational areas, refuse disposal traffic congestion, etc. The rivers passing through or passing by the side of almost all the cities are grossly polluted. 57% of the population of class I cities and 80% of class II cities in India are answered. Pollution of water courses in urban areas are mainly due to sewage, sullage, solid waste leachates and industrial wastes.

The population of India was 236 million in 1891, and it was around 472 million in 1964, and in other words, doubling time of the population was 73 years. But after 1951, as the process of development gathered momentum there was sharp decline in death rate, accompanied by a very slow and lagged decline in birth rate. As a result of it, the country is experiencing a phase of population explosion and by AD 2000 the population of the country will touch I billion mark. The figures obtained in 1991 census revealed that all population forecasts have been belied. According to eminent demographers, Indian population was more than 800 million in 1991 and the growth of population during 1971-91 was of the order of 245 million, i.e. more than the total population of the country in 1891. This extraordinary high rate of population growth has its impact on development and environment. It is now generally recognised that population, resources, environment and development are inextricably complex. These interrelationships need to be understood for proper appreciation of the issues involved in the integration of environment with development.

Along with the increase in general population in the country, urbanisation and industrialisation is also on increase. Indian cities have been growing at an alarming rate over the past few decades. Against a national population growth rate of 2.1-2.5 per cent, Indian dries with population above 1 million have grown at an annual growth rate of 33-45 per cent. In 1991, over 38 per cent of the Indian population was dwelling in the urban areas. At this rate die population living in urban areas in India will increase to 45 to 50 per cent by AD 2000.

This trend of urbanisation is a natural corollary of the process of economic growth through industrialisation being followed in India. What is very peculiar in the Indian context is that the growth of the larger cities have become unmanagable. Large cities growing at the rate higher than national growth rate has meant depletion in the population of smaller towns and denuadation of country side.

The population of greater Bombay which was 1.8 million by the year 1941 had increased to 125 million by the year 1991. Continuous increase in population without commensurate extention in Bombay's Landmass resulted in increase in congestion-Deterioration quality of life and dwindling open spaces brought a few concerned citizens together and that is how the Save Bombay Committee was formed in 1972.

The population of Calcutta increased from 55 million in 1960 to 10.6 million in 1990, while the population of New Delhi increased from 23 million in 1960 to 9.44 million. The population of Madras increased from 1.7 million in 1960 to 8.4 million in 1990. Same is the case in Hyderabad, Bangalore, Kanpur, Ahmedabad, Trivandrum, etc. The expected population by AD 2000 in Bombay, Calcutta, Madras and New Delhi are 17.1 million, 16.7 million, 10.9 million and 112 million respectively. From these figures it is clear that the population of dries in India which has doubled during 1970-1990 is likely to double again during 1990-2000. The evils of agglomeration and urban concentration are many and varied. With the skewed urban development which characterised India during the past three decades, the environment has deteriorated visibly and considerably. The urban areas are suffering from their own blight: squatter settlements, lack of sanitation and water supply, over-crowding, congestion and pollution.

Urbanisation and the Environment

The process of urbanisation generally proceeds the process of industrialisation. As such it continually lacks the supporting economic potential. As industrialization tries to catch up with urbanisation process, the lag between the two trends to become larger and longer. As such, the problem becomes more severe, resources become more scarce and organisational inadequacy more stark. Normal and disguised unemployment becomes and migrants tend to enter the service industry which is not productive. Crime and delinquency become more and more pronounced. As the urbanisation process accelerates, energy consumption tends to increase. The per capita rate of energy consumption in general increase when the people migrate from rural to urban areas. What is more the per capita consume person of non-renewable energy increase due to shift use from firewood to coal and from vegetable oil to kerosene and electricity.

The large cities in India like Calcutta, Bombay, Delhi, Madras, Hyderabad, etc., are thus facing the traditional environmental problems such as lack of sanitation, water supply, chronic shortage of services, polluted air and water of open space and recreational areas, refuse and solid waste disposal, traffic congestion etc. As an example of overcrowding; more than 20 per cent of urban households in India occupy a single room of 100 sq.ft. or less with an average occupancy of 4.5 persons per room. Nearly 50% of the households have less than 200 sq.ft. and 75% contain less than 300 sq.ft. Over one-third of the housing units are temporary structures. Lack of overall environmental standards of fresh air open space and basic services for health and hygiene are converting the existing houses into neo urban slums. Housing space is reduced often by the inflow of unemployed relatives and other new migrants. The situation is further aggravated by the gradual rise in building costs. As a result every year greater number of families are added to the list of the people without shelter. The number of families living on footpaths is increasing day by day.

About one-third of the urban population in India is believed to live in squatter settlements. The disabilities from which the squatter settlements suffer are (a) Land tenure is frequently insecure and haphazard (b) The residents suffer from health problems particularly gastrointestinal and respiratory diseases, both related to environments conditions and to the heavy demand for shelter. Community organisation is based on kinship or place of origin, (c) Low income and weak purchasing power prevail, (d) Shelter is generally substandard as a result of poverty, insecurity of land tenure and household investment priorities, (e) Public utilities such as water supply, sanitation, electricity, transport etc, are insufficient and inadequate.

Thus, the urban areas in India face the critical and complex environmental problems which adversely affect the carrying capacity of ecosystem to sustain such giant conurbations at existing levels of growth and consumption. The rivers in passing through or passing by urban areas in India are grossly polluted. The estuaries or marine environments near urban settlements are contaminated needing immediate attention. The unhygeinic disposal of solid wastes from residences, commercial undertakings, industries, hospitals are creating numerous problems including water pollution due to leachates from solid waste dumps.

Waste water disposal is a problem in itself for urban communities. Though the larger public sector factories have often there own waste water disposal system there are many industries with no such facilities. Waste water causes water pollution, obnoxious odours and air pollution, harbours diseases, vectors such as insects and contributes to other health hazards.

Water Supply and Sanitation in Urban Areas

Few needs are more basic than those of good water and proper sanitation. It is no exaggeration to say that public water supply is the foundation on which rests the economics of the community. Water supply is gaining in importance as a location factor and is playing a decisive role in the location as an increasing of new or expanding industries and even in relocation of existing ones representing the problems of inadequate quality of water and rising cost. In addition, water is required for agricultural use, hydroelectric power generation, fisheries, drinking, navigation, etc. All of which are related to the economy of the community. A community with a restricted water supply is therefore, a community with restricted growth.

There is no truer sign of civilization and culture than good sanitation. A good drain reflects the culture as much as a beautiful statue. The accomplishments of good sanitation are not reflected alone by statistics of sickness of death traceable to water-borne infections. It is intimately connected with the quality of life. Typhoid fever, a scourage at the beginning of the present century, exacted in North America an annual death toll of more than 115 per 100,000 population, has been controlled by proper methods of water supply filtration and disinfection in particular and by other sanitary safeguards, such as safe methods of waste water disposal, milk and food supplies and control of flies and mosquitoes. Cholera, which was pandemic during the nineteenth century, is confined at present principally to the Asiatic continent.

In India many cities like Madras, Bangalore, Hyderabad, Calcutta, etc., have acute shortage of water. These cities are forced to obtain water from a source far away from them. For example, in Hyderabad, Osmansagar, Himayat sagar reservoirs and the Manjira which constitute the principal source of drinking water is over exploited. For the first time the twin cities are supplied water in 1986 on alternate days. The Hyderabad Metro water works are allotted 200 million rupees in the 1986-1987 budget to expedite tapping of water from Singur project. Andhra Pradesh Government has plans to seek World Bank Aid to bring Godavari water, to Hyderabad from a long distance. Madras city plans to get water from Krishna river from a far away place through Telugu-Ganga Project. Water received from Cauvery has become inadequate in Bangalore. The explosive growth of population coupled with unplanned urban migration rendered existing public water supplies in most of the cities in India inadequate. Their expansion became imperative yet sources of financing are not readily available.

There has been considerable increase in the provision of urban water supply in India during the last 20 years. The targets in Water Supply and Sanitation Decade ending in 1990 is to provide safe water to 77% of the urban population in India. Lack of political will, inadequate and inappropriate organisational and administrative set up, lack of trained personnel at appropriate levels and inadequate community participation have all contributed to the slow progress in this vital sector.

Better water and improved sanitation together lead to better health but improvement of water supplies alone without concomitant provision of sanitary disposal of excrete is much less effective. The emphasis on 'clean water' should not lead to neglect of the equally urgent need for sanitation. More water supplies will generate more waste water. Indiscriminate disposal of this waste water will lead to problems of mosquito breeding and consequent insanitary conditions. Mosquitoes are on increase in urban centres in India due to insanitation conditions. For examples, the worst nuisance for citizens of Hyderabad and other cities mosquitoes.

A WHO survey of 61 developing countries showed that only 27% of the urban population was connected to public sewers, 43% relying in individual household systems and the rest without any facility. Increase in the facility of public sewerage system is not keeping pace with the increase in urban population. Even more serious is the fact that 30% of population continue to have no access to any sanitary facility at all. In India, by 1980, 27% of the urban population is covered by public sewerage. The target for 1990 is 80%. The status of (population 50,000 to 100,000) India is shown in Table 6.6.

Table 6.6. Status of Sewerage System in Urban Areas in India

	*Class I Cities**	*Class II Cities"*
Number	142	198
Sewered population	43	12
Unsewered population (%)	57	8S
Waste water generated (in MLD)	7000	1200
Waste water collected in MLD	4300	190
Waste water treated in MLD	2700	66

Class I cities population above 100,000

** Class II cities population between 50,000 and 100,000.

Lack of skilled manpower, inadequate capital, inappropriate technology, insufficient community participation, outmoded institutional infrastructure for programme planning and implementation and shortage of managerial skill are coming in the way of provision of basic sanitary services in India.

Supply of clean water and adequate sanitation facilities is an essential requirements in promoting environmental healter, which is weak in this country.

Refuse Disposal in Urban Areas in India

Solid waste disposal is assuming serious dimensions in the urban areas in India. The amount of solid waste generated per capita in India is nearly four times that of dry weight of sewage solids. But the amount of money and effort spent on solving this problem is comparatively meagre.

In India primitive methods of solid waste disposal have been practiced in many cities hitherto. These have been highly unhygienic and unsatisfactory involving manual handling of wastes in loading, unloading and landfilling or composting operations.

India can take some credit for developing composting methods for urban areas, for example, the "Bangalore method" and the "Indore method" which were first developed in India in 1928. While town and medium sized cities in general adopted composting with success, it is in the big cities that several problems are posed like costs, land availability and marketing of the compost. Generally, in India, the alternative for composting has been plain dumping of solid wastes in low lying areas. Environmental pollution problems are caused by such dumping of solid wastes. In Bombay, the dumping of refuse in low lying areas adjoining the creeks has led to severe pollution of creek water where BOD values as hit as 400 mg/1 have been recorded. Instances have been observed of migratory birds avoiding these creek areas. Besides water pollution, such wastes also undergo spontaneous combustion giving localised air pollution problems, and harbour vectors of disease such as flies, rats etc. Sanitary land filling practices which avoid pollution have been conspicuous by their absence. Absolute land of social hygienic and highly unsatisfactory systems of refuse storage, collection and disposal are creating health hazards and pollution problem in all Indian cities, but nowhere more so than in Calcutta. Even near the Jagat Cinema Hall adjacent to Jama Masjid in Delhi, people live just by the side of a massive municipal refuse collection centre. No need to describe the cities of Bihar, Orissa, West Bengal, etc.

The per capita solid waste reaching disposal sites in Bombay, Calcutta, Madras and New Delhi range from 0.5 to 0.6 kg per day. In Hyderabad the per capita value is 0.33 kg/day. In other Indian cities the per capita contribution ranged from 0.15 to 0.33 kg/day.

Density value of refuse in Indian cities varies from 300 to 560 kg/m. Composition of refuse from different cities in India is shown in Table 6.7.

Table 6.7 Characteristics of Refuse Fronable Cities in India

(All values in % by wet weight)

Characteristics	*Calcutta*	*Delhi*	*Madras*	*Hyderabad*
Paper	3.18	6.29	7.85	4.81
Plastics	0.65	0.85	0.99	0.83
Metals	0.66	1.21	0.95	1.22
Glass	0.38	0.57	0.96	0.93
Ash and buries Compontible	34.00	36.00	28.00	36.00
Total				
matter	47.00	35.00	48.00	37.00
C%	19.20	-	21.48	-
N%	0.56	0.76	0.68	0.61
P%	0.59	0.69	0.56	0.49
K%	0.43	0.46	0.48	0.45

One of the major problems with the disposal of these solid wastes has been water pollution from land fills. A familiar scene in many of the cities in India is over flowing refuse bins in several locations and uncleared refuse heaps at street corners. For example, 23 million people of Hyderabad and Secunderabad generate 750 tonnes of refuse per day and the corporation has nearly 5000 women sweepers, workers and scavengers. With the pollution increasing and slum proliferating conservancy has become a major problem as there is no appreciable increase in the strength of staff. A fleet of 135 trucks, tractors, hippo haulers pick up refuse from the 6500 bins and 127 refuse transfer stations and dump it at the trenching grounds on the out skirts. With inadequacy of fleet on any given day one-third of refuse bins remain uncleared.

Solid wastes generated from industrial sources are heterogenous ranging from inert inorganics as in those produced in mining, collieries, to organics from those producing basic consumer products, and may include even hazardous wastes as in nuclear industry. The solid wastes from industries can be grouped as (a) biodegradable (b) non-biodegradable (c) hazardous biodegradable solid wastes age generated in industries like fruit processing, canning, slaughter house, cotton ginning, textile, etc. Non-biodegradable wastes are generated in industries like refineries, collieries, thermal power plants, paper industry, ferrous and nonferrous metallurgical industries, etc. Solid wastes generated from urban and industrial sources contain large number of ingredients, some of which are toxic or hazardous. Bulk of these toxic or hazardous wastes containing Arsenic, Cyanide, Mercury, Chlorinated hydrocarbons, etc., are generated in the industries.

In the urban areas in India there are many industries like refineries, thermal power plants, electroplating shops, fertilizer factories, etc. They all generate solid wastes of different nature. The impact of their disposal on environment in general and water pollution in particular in urban areas is not at all well studied.

Air Pollution in urban Centres and Consequent Water Pollution

According to the second citizens Report (1984-1985) on the state of India's environment the quantity of sulphur dioxide released into the air has tripled in the past 15 years in India. Of the 48 thermal power stations officially surveyed in 1984, 31 had taken no pollution control measures only 6 had their equipment functioning properly. Coal output in India has increased from 35 million tonnes in 1950 to 150 million tonnes in 1980. It is reported that India produces every year 4 million tonnes of Sulphur dioxide, 7 million tonnes of particulates, 0.5 million tonnes of oxides of nitrogen, 0.5 million tonnes of carbon monoxide wet dispositions found to be quite predominant particularly in Delhi, Pune and Nagpur, etc.

Energy consumption, automobiles and industries together have vitiated the urban atmosphere. Calcutta, Burdwan, Bombay, Pune, Thane, Coimbatore, Madurai, Ernakulum and Kanpur account for 80% of the air pollution in the country. Bombay with its two refineries, 2 thermal plants and a number of chemical and fertilizer plants emits 350-400 tonnes of Sulphur dioxide everyday. Some of industrialised urban areas like Chembur in Bombay, Basin bridge area in Madras, Indraprastha estate in Delhi, Howrah bridge area in Calcutta and Udyog mandal area in Kerala, has attracted the attention of public, Industries as well as Government with respect to air pollution and its consequence to water pollution with setting dust.

Vehicular traffic is the main source of acid pollution in Indian cities. Already Delhi has the country's worst air pollution problem due to heavy traffic. Air pollution survey has revealed that 400 tonnes of pollutants are emitted every day by nearly 500,000 vehicles. Two and three wheelers constitute 63% of the total vehicles now plying on the Delhi Road, "Work carried out by Central Water Pollution Control Board showed that the exhaust gas from two-wheelers, three-wheelers and four-wheelers contained in volume basis 4.12%, 3.86% and 6.63% of carbon monoxide respectively. Average density of smoke in Hartridge units from Tata, Ashok Ley land and other trucks was 94.44,92.86 and 91.50 respectively. Dust emission from Indraprastha Thermal power station from units 3 and 4 was 28.9. and 27.32 MT/day. Acid pollution in the form of dry deposition has caused problems for Indian Railways. Acidic fumes, dust and moisture have affected the high tension insulators along electrical railways tracks. Extensive damages to forests, crops, lakes and aquatic life have not yet become too obvious in India, either because of most of the pollution industries are located in the vicinity of urban centres or such damages have not been linked to acid pollution so far.

India today enjoys the distinction of being the tenth most industrialised country in the world. Most of the industrialisation has been achieved in the post-independence period at rapid rate. Further, the range of industrialisation has been very wide covering even the most sophisticated fields such as nuclear power and space research.

Such a fast and wide-ranging industrialisation has brought into its wake numerous hazards to workers working inside the factory and the people living outside the industry. In December 1984, Bhopal witnessed the unprecedent tragedy caused by massive leakage of methyl isocyante and its coexisting traces of phosgene gas from the Union Carbide pesticide plant. Apart from killing 2000 people, blinding several thousand and affecting over 150,000 people, a degree of foetus damage in the case of pregnant women and deep-seated psychological trauma city-wide has been reported. Later ammonia leakage in Bombay, Oleum leakage in Delhi, Chlorine leakage in Madras and other places are reported.

Although chemicals have benefited human society in many ways, their widespread use began to cause concern for the environment and human health in recent years. Studies indicated that radium originally used on watch faces for luminous figuring caused cancer. Similarly dibromo-chloropropane originally was direct. Vinyl chloride monomer caused angiosarcoma of the liver. Accidents during the transportation of chemical in bulk could require, evacuation of whole communities. One of the greatest perils to human life is the relatively persistent toxic chemicals and metals which can build upto dangerous levels before being detected. The mercury pollution in Minimata Bay in Japan began in 1939 but actual cause of Minimata disease was realised in 1973. Similarly for over fifty years, only recently they have been found to cause cancer, disease, jaundice and liver damage. The consequence of air pollution has a linkage with water pollution.

Waste Water Disposal in Urban Centres

The survey conducted by the Central Water Pollution Control Board suggested.

(i) The total industrial waste water generated in the country constituted only 10 per cent by volume of the total waste water generated by all urban settlements.

(ii) The cottage and small-scale sector together contributes only 30% of the total industrial waste water and hence it is only 3% by volume of total waste water generated by all urban settlements.

(iii) The industrial waste water contributed by large and medium sector industry constitutes the remaining 7% by volume of total waste water generated by all urban settlements.

The above listed facts establishes that control of domestic waste water from at least Class I cities must receive the priority it deserves. The preliminary estimate by Central Water Pollution Control Board suggests a requirement of Rs. 18000 million which may be planned for 3 plan periods providing Rs. 6000 million in each plan period. As suggested by central water pollution control board the phasing could be as follows:

(i) *Phase 1 Rs. 6000 million.* Twelve metropolitan cities (population 1 million and above). Ahmedabad, Bangalore, Bombay, Calcutta, Delhi, Hyderabad, Jaipur, Kanpur, Lucknow, Madras, Nagpur and Pune.

(ii) *Phase 2 Rs. 6000 million.* Class I cities having half the population sewered and thereby damaging water environment through sewer out fall should be taken in phase 2.

(iii) *Phase 3 Rs. 6000 million.* Class I cities having no sewage and no sewage treatment plants.

The progress in abating pollution from domestic sewage originating from human settlements is insignificant. The progress in sewerage sector even in Class I cities and Class II towns is very low. It is needless to mention that the success of Water Act is forgery dependent on a strong implementation programme of hygienic disposal of sewage emanating from human settlements. In this regard the target fixed by the Ministry of Housing and Works, Government of India for 100 per cent coverage of population for Class I cities with sewerage and sewage treatment facilities and low cost sanitation methods in respect of other towns during the Decade period (1981-91) is a significant programme. Successful completion of this programme would improve the water quality in receiving bodies, taken seriously?

All waste waters, domestic or industrial, ultimately manage to get into riverine systems of the country. Although the sea is the ultimate sink but river plays the major role of transport.

Water quality monitoring in inter-state rivers by Central Water Pollution Control Board showed that the locations where water quality is found are downstream of Ahmedabad on the Sabarmati and down stream of Pune on the Bhima, a tributary to the Krishna. The survey of water quality in Ganges river by Central Water Pollution Control Board showed polluted stretches in the over below Kannauj, Kanpur, Allahabad, Varanasi, Patna and Calcutta. It proves that city sewage in all the polluted stretches is the major cause for such reachwise pollution.

The Musi in Hyderabad is so polluted that the Andhra Pradesh State Pollution Control Board has categorised it as 'below class' meaning its water is fit for controlled waste disposal only and irrigation. The city's sewage and sullage, is let into it even as the surplus nala of Hussainsagar lake carries sewage and sullage, to join the Musi, Wet cultivation of paragrass in the river bed and neighbourhood and stagnant sewage water provide deal ground for mosquito breeding. After a decade of effort, legislation to ban the cultivation within the city was enacted recently.

The Cooum river, Adyar River and Buckingham Canal passing through Madras are highly polluted mainly from sewage and sullage, and industrial wastes. The Valgai river near Madurai is grossly polluted.

The Yamuna river between Delhi and Agra is so much polluted that it has drawn the attention of one and all. The Mahi, the Narmada, the Tapati, the Krishna, the Pennar, the Godavari and Suvamarekha are also seriously polluted in different stretches.

Organic and biological pollution in the inland and estuarine waters near the urban centres, industrial and port cities is mainly due to dumping of latfreated or partially treated domestic wastes into the nearest water area or into the sea.

Industrial estates housing a number of small, medium and cottage industries are posing a problem of water pollution in India. Because of the lack of appreciation on the part of the promoter of the collective impact of many such medium, small and cottage units on the environment, the promoting agencies have done enough damages to the environment and water environment in particular. In majority of the estates in the country no facilities for sewerage and terminal waste treatment plants was conceived. The Central Board initiated a case study at Vapi industrial estate. Gujarat, the largest estate in the country, demonstrated how co-operative collection, treatment and disposal of such individual wastes of the units in the Estates are possible. Similar studies were done by State Pollution Control Boards in the country and some of the Estates have started combined waste water treatment systems.

Central as well as 14 State Water Pollution Boards conducted an Industrial survey during April-May, 1981, to identify water polluting industries in the large and medium sector industries. Out of a total 27,000 large and medium industries 1700 water polluting industries were identified. About half of them are in urban areas. For example in Kanpur, Bombay, Ahmedabad, Bangalore, Madras, Calcutta, etc. There are about 100 industries each causing water pollution in urban areas. As per the survey by 1991 there were 460 industrial effluent treatment plants under operation or commissioning stage and 251 treatment plants either under construction or under planning.

Central Water Pollution Board prepared an Industry Document for each of the Specific Industry as shown in Table 6.8.

Table 6.8. Industry-specific Pollution Control in India as in 1992-1993

	Typc of Industry	*No of units*	*Effluent treatment plant status Operation or Commi-ssioning*	*Cons-truction or planning*	*Total*
1.	Sugar	300	86	18	104
2.	Distillery	128	28	6	34
3.	Caustic soda	38	21	6	27
4.	Fertilizer	65	48	7	55
5.	Oil refinery	13	13	Nil	13
6.	Man-made fibre	29	16	6	22
7.	Integrated steel plant	7	5	2	7
8.	Textile (wool and cotton)	300	125	30	155
9.	Pulp and paper	150	10	45	55
10.	Pharmaceuticals	120	28	28	56
11.	Pesticides	50	12	4	16
12.	Petrochemicals	10	6	4	10
13.	Organic chemicals	150	4	30	34
14.	Dairies	50	20	12	32
15.	Thermal power plants	150	30	29	59
16.	Non-ferrous metals	12	4	4	8
17.	Dye manufacturing, Large, Engg. using Electroplating, Large tanneries and Misc tanneries	128	4	20	24
	Total	**1700**	**460**	**251**	**711**

Such remains to be done in controlling pollution from medium as well as large scale industries in urban as well as rural areas in India. Waste water treatment is essentially a management problem. More often than not this management problem could result in profits for the industry whilst fulfilling its commitment to environment. Total waste water management approach calls for waste water volume reduction, waste strength reduction byproduct recovery and evaluation of various treatment units. Water is a misplaced resource and the challenge in waste management is to identity a scheme of modifying inputs, modifying production processes and recovering byproducts to result in profits for the industry while satisfying effluent standards stipulated by Water Pollution Control Boards.

The possibility of developing fundamentally new processes for waste water treatment is remote. Only an innovative use of existing technology to arrive at functional and cost-effective management scheme best suited to local conditions would afford the possibility of a productive proposition. There are situations where non-availability of water forced curtailment of industrial production in summer. A waste management scheme resulting initial investment in treatment that could be recovered in due course of time. Then there are industries where land is scarce and only a recover to physico-chemical treatment would produce an effluent acceptable to the Pollution Control Board while satisfying land is available simple biological processes exploiting the natural advantages of bright sunshine and high temperature with a minimum requirement of equipment, energy and skilled manpower would result in an appropriate and reliable treatment system. The regional waste water treatment plants jointly treating municipal and industrial effluents or combined treatment and effluents from several industries generally result not only in a less expensive solution due to economy of scale concomittant neutralisation, equalisation and proportioning by also in a cost sharing financial system based on "pollution pays" principle.

Pic. 6.1 From a spring to a raging mass of water

Pic. 6.2 For a greener planet : Be aware of your environment

Pic. 6.3 Waiting to Sink : Tehri town, soon to be submerged in the reservoir of the Tehri dam, in Uttaranchal on Tuesday. Around 125 villages and more than 100,000 people have been shifted and rehabilitated in a controversial process. The Tehri Dam power project is expected to produce 1,000 MW when it becomes operational next year.

7

AIR POLLUTION

In recent years, great concern has been universally voiced regarding environmental pollution arising as a side effect of industrial activities. In our national situation with the growth of industries, more and more toxic substances are either used as raw materials or given off during manufacturing processes in the form of dust, fumes, vapours and gases. These pollutants ultimately dissipate in the working environment and pose occupational health hazards. According to a U.S. Estimate, there are some half a million man-made substances already present as pollutants in our total environment. These pollutants, when get into the air and water disturb the ecological balance of nature. Control of air contaminants is therefore essential for, we are just the custodians of environment for our future generation . This can be achieved by properly recognising, evaluating and controlling the industrial environment. Considering the above, we in India, may well be satisfied with our consciousness in this matter and it may be mentioned that several Air and Environmental Authorities have been in existence from the 1960-65 period onwards. An Indian 'Water Pollution Bill' came into force from 1969 and a National Committee on Environmental behaviour started functioning from the early seventies. As a result of increasing awareness, four more acts came into existence which are the

Water (prevention and control of Pollution) Act, 1974, the Air (prevention and control of pollution) Cess Act 1977, the Air (prevention and control of pollution) Act, 1981 and the Environment Protection Act, 1986.

The dry air contains 78.09% nitrogen by volume and 20.94% oxygen, the remaining 0.97% is composed of gaseous mixture of carbon dioxide (0.03%), helium, argon, krypton xenon, neon (i.e. inert gases) nitrous oxide and water vapour (varying amount) as well as very small amounts of some other organic and inorganic gases whose amount in the atmosphere vary with time and place. Various amounts of contaminants continuously enter the atmosphere through both natural and man-made processes that exist upon the earth. That portion of these substances which interacts with the environment to cause toxicity, disease, aesthetic distress, physiological effects or environmental decay, has been labelled by man as a 'pollutant'.

In general, the actions of people are the primary cause of pollution and as the population increases, the pollution problems also increases proportionately. The first significant change in man's effect on nature came with his discovery of fire. Prehistoric man built a fire in his cave for cooking, heating and to provide light. The problem of air pollution came into existence at this time.

The British Parliament passed an Act in 1273, forbidding the burning of coal in London because it was beginning to choke the atmosphere. In 1300, King Edward I issued a royal proclamation, "Whosoever shall be found guilty of burning coal all suffer the loss of his head." In 1306, a man was executed for violating this regulation. Later, the law fell into disuse as the industrial revolution took place in England.

Today, the natural terrain that surrounds large cities is recognized as having a significant beaming on the problem of air pollution. However, this is not an altogether new concept either. Historians tell us that Los Angeles which in recent years has become a national symbol of comparison for excessive smog levels, was known as the 'valley of smokes' when the Saniards first arrived there.

It has been found that a significantly increasing volume of particulate matter entering the atmosphere, scatters the incoming sunlight. This reduces the amount of heat that reaches the earth and tends to reduce its temperature. This decreasing the mean global temperature in recent years have been attributed to concentration of air-borne particles in the atmosphere. A counter acting phenomenon commonly referred to as the "greenhouse effect" is caused by the increasing amount of carbon dioxide found in the atmosphere. It has been estimated that if carbon dioxide content in the atmosphere generated in combustion processes continues to increase at the present rate, the mean global temperature could rise by 4°C in the next five decades. There has been conjecture that this might become a matter of great importance because small temperature increase could cause a partial melting of the ice caps of the earth causing continental flooding and devastating effects on man.

Air pollution can cause death, impair health, reduce visibility, bring about vast economic losses and contribute to the general deterioration of both our cities and country-side. It can also cause intangible losses to historical monuments such as the Taj Mahal which is believed to be badly affected by air pollution.

On account of large scale industrial activities in advanced countries notably the USA, UK, and other European countries, fall of acid rain has been reported in Scandinavia. This has reduced forest growth in Scandinavia. In Canada, thousands of lakes have been destroyed due to acid rain. Apart from the international issues involved, the basic ecology is affected. Large scale deforestation apart from creating an imbalance in the oxygen proportion of the atmosphere, affects weather and rain patterns as well. Industrial activity, particularly in thermal power stations, cement plants, oil refineries, chemical complexes, metallurgical industries, steel plants and fertilizer complexes, causes major problems of air pollution. The effects of air pollution are felt more by the elderly and infants as chest and respiratory complaints are very common among them. Tragic instances of death have also been reported in air pollution episodes such as the London smog of 1952, and the Bhopal gas tragedy of 1984.

It is, therefore, a matter of great importance that engineers of all disciplines consciously incorporate in their designs sufficient constraints and safeguards to ensure that they do not contribute to atmospheric pollution, in addition, they must apply their ingenuity and problem solving abilities to eliminate air pollution where it exists and restoring the natural environment.

There are three methods of identifying air pollution:

1. Sensory recognition.
2. Physical measurements of pollution.
3. Effects on plants, animals, and buildings.

These methods are widely used and they have their merits and demerits.

1. *Sensory recognition*-Usually the first awareness of an air pollution problems is through some effects on the individual. These are:
   (a) strong or unusual odours.
   (b) Reduction in visibility.
   (c) Eye irritation.
   (d) Acid taste in the mouth.
   (e) Feel of grit under foot.

There are highly subjective phenomena and vary from individual.

(2) *Physical measurement*-While sensory perception may provide the first indication of the presence of most of the contaminants in the air, it is often not possible to detect or trace quantities of many air-borne toxic substances or the presence of radioactive matter through the senses. Their identification required

physical measurement by standard methods of sampling and analysis.

(3) *Effects on plants, animals and buildings*-Effects of air pollution can be observed on the growth of plants and health of animals. Similarly, its deleterious effect on building can also be observed. Thus plants, animals, and buildings act to some extent as indicators of certain atmospheric impurities.

Classification of pollutants

Air pollutants can be classified as follows:

1. Natural contaminants

   e.g., natural fog, pollen grains, bacteria, and products of volcanic eruption

2. Aerosols (particulates)

   e.g., dust, smoke, mists, fog and fume

3. Gases and vapours

The various gases and vapours, which are important air contaminants are given in Table 7.0.

Table 7.0: Air Contaminants

No.	*Group*	*Examples*
1.	Sulphur compound	SO_2, SO_3, H_2S, raercaptans
2.	Nitrogen compounds	NO, NO_2, NH_3
3.	Oxygen compounds	O_3, CO, CO_2
4.	Halogen compounds	HF, HCI
5.	Organic compounds	Aldehydes, hydrocarbons
6.	Radioactive compounds	Radioactive gases

In addition to the above, studies by various investigators have indicated that some of these contaminants undergo chemical reactions when they enter the atmosphere. As a result the end products formed are more harmful than the original contaminants. For example, it is reported that unsaturated hydrocarbons react with nitrogen dioxide in sunlight to form smog.

Natural Contaminants

Among natural contaminants pollen is important because of its peculiar properties irritating to some individuals. Pollen grains are the male gametophytes of gymnosperms and angiosperms and they are discharged into the atmosphere from weeds, grasses and trees. Because of wind pollination, thousands of pollen grains are liberated. While air transported pollen grains range chiefly between 10g and 50g (micron) in size, some have been found to be as small as 5g and as large as 100g in diameter.

From the point of view of pollution, air pollutants are significant because of the allergic responses produced in sensitive individuals. Many people suffer from asthma or high fever. While most victims have an uncomplicated type of hay fever in which the symptoms disappear at the end of the pollen season, some develop bronchitis, bronchial asthma, and dermatitis.

Aerosols

Aerosols refer to the dispersion of solid or liquid particles of microscopic size in gaseous media, such as dust, smoke, or mist.

An aerosol can also be defined as a colloidal system in which the dispersion medium is a gas and the disappeared phase is solid or liquid.

The term 'aerosol' is used during the time it is suspended in the air. After it has settled either by virtue of its weight, by agglomeration, or by impact on a solid or liquid surface, the term no longer applies. Thus, particulate matter is an air pollutant only when it is an aerosol. However, it is a nuisance both as an aerosol (visibility reduction) and as settled or deposited matter (soiling of surfaces, corrosions).

Aerosols differ widely in terms of particle size, particle density and their importance as pollutants. Their diameters generally range from 0.01g or less, up to about 100g. The following are the various aerosols.

Dust

Dust is made up of solid particles predominantly larger than those found in colloids and capable of temporary suspension in air or other gases. They do not tend to flocculate except under electrostatic forces; they also do not diffuse but settle under the influence of gravity.

Dust is produced by the crushing, grinding etc., of organic and inorganic materials. Generally they are over in diameter, although some are smaller. Fly ash from chimneys varies from 80-3μ; cement, from 150-10p; foundry dust, from 2C3-lp. Most of the dust particles settle to the ground; 5 dust fall, but particles 5μ or smaller tend to form stable suspensions.

Smoke

Smoke consists of finely divided particles produced by incomplete combustion. It consists predominantly of carbon particles and other combustible material. Generally the density of the particles is less than 1μ. The size of coal smoke particles range from 0.2-0.01p and oil smoke particles from 1.0-0.03μ.

Mists

This term refers to a low concentration dispersion of liquid particles of large size. In meteorology, it means a light dispersion of minute water droplets suspended in the atmosphere. Natural mist particles formed from water vapour in the atmosphere are rather large, ranging from 500-40 1 j in size. The particles may coalesce.

Fog

Fog refers to visible aerosols in which the dispersed phase is liquid. Formation by condensation is usually implied. In meteorology, it refers to dispersion of water or ice in the atmosphere near the earth's surface reducing visibility to less than 1/2 km. In natural fog the size of the particles ranges from 40-1.0μ.

Fumes

These are solid particles generated by condensation from the gaseous state generally after volatilization from melted substances, and often accompanied by a chemical reaction such as oxidation. Fumes flocculate and sometimes coalesce.

Table 7.1 indicates the major sources of atmospheric dust.

Table 7.1: Sources of Atmosphere Dust

Sources	Examples
1. Combustion	Fuel burning (coal, wood, fuel oil) incineration (house and municipal garbage) Others (open fires, forest fires, tobacco smoking)
2. Materials handling and processing	Loading and unloading (sand, gravel, coal, ores, lime, cement) Crushing and grinding (ores, stone, cement, rocks, chemicals) Mixing and packaging (chemicals, fertilisers) Food processing (Hour, cornstarch, grains) Cutting and forming (saw mills, wall board, plastics) Metallurgical (foundries, smelters) Industrial paper, textiles manufacture)
3. Earth-moving operations	Construction (road, buildings, dams, site clearance) Mining (blasting) Agriculture (soil filling, land preparation) Winds
4. Miscellaneous	House cleaning, Mud road cleaning, Crop spraying. Poultry feeding Engine, exhaust

Gases

Following are the gases in air pollutants:

Sulphur Dioxide

This is one of the principal constituents of air pollutants. The main source of sulphur dioxide is the combustion of fuels, especially coal. Therefore, its concentration in the atmosphere depends upon the sulphur content of the fuel used for heating and power generation. The sulphur content of fuels varies from less than 1% for good quality anthracite to over 4% fur bituminous coal. In the USA in recent years, there has been a progressive decrease in the average atmospheric concentrations of sulphur dioxide, of several cities, owing to the increased use of coal with a low sulphur content. Air pollution control measures in St Louis prohibit the use of high-sulphur coal.

Most crude petroleum products contain less than 1% sulphur: a few contain upto 5%. Refining processes tend to concentrate sulphur compounds in the heavier fractions. Fuel gases also contain sulphur, but, in small quantities.

About 80% of the sulphur in coal and nearly all that in liquid and gaseous fuels is found in flue gases in the form of sulphur dioxide. The remaining sulphur in coal is present as inorganic sulphur and thus remains in the ash. Generally, the concentration of sulphur dioxide in flue gases ranges from 0.005-0.25% and occasionally is as high as 0.4%.

Another common source or sulphur dioxide in the atmosphere, is metallurgical operations. Many ores, like zinc, copper and lead, are primarily sulphides. During the smelting of these ores, sulphur dioxide is evolved in stack concentrations of 5-10% (SO_2) but this, can be recovered in the form of sulphuric acid.

Among the miscellaneous operations releasing sulphur dioxide into the atmosphere are sulphuric acid plants and paper manufacturing plants. The quantities are usually low and therefore easily amenable to control measures.

The open burning of refuse and municipal incinerators also contribute some amount of sulphur dioxide to the atmosphere.

Hydrogen Sulphide

Hydrogen sulphide is a foul smelling gas. The sources of its natural emission include anaerobic decay processes on land, in marshes and in the oceans. Volcanoes and natural water springs emit hydrogen sulphide to some extent.

One of the major sources of Hydrogen sulphide is the Kraft pulp industry, which uses a sulphide process for manufacturing paper. The other industrial sources of hydrogen sulphide are petroleum refineries, coke-oven plants, viscose rayon plants, and some chemical operations.

Other sulphur compounds that are of interest in air pollution, principally because of their strong odour, are methyl mercaptan (CH_3 SH), dimethyl sulphide (CH_3 SCH_3) dimethyl disulphide (CH_3 SS CH_3) and their higher molecular homologs. The mercaptans are emitted in mixtures of pollutants from some pulp mills, petroleum refineries, and chemical manufacturing plants.

Hydrogen Fluoride

The major sources of fluorides are the manufacturers of phosphate fertilizers, the aluminium industry, brick plants, pottery, and ferro-enamel works. Small amounts are also emitted from other metallurgical operations, such as zinc foundries and open-hearth steel furnaces. Small amounts are also liberated in the burning of coal, which normally contains about 0.01% fluorine.

Hydrogen Fluoride is an important air contaminant even in extremely low concentrations of 0.001-0.10 ppm by volume. In these concentration levels, hydrogen fluoride is more important in terms of injury to vegetation and animals than in terms of injury to humans. The high degree of toxicity of fluorine compounds renders the control of such emissions imperative for industries manufacturing aluminum and phosphate fertilizers.

Chlorine and Hydrogen Chloride

Chlorine is found in polluted atmospheres as the element itself (chlorine) as hydrogen chloride, as chlorine containing organic compounds such as per chloroethylene and as inorganic chlorides. The last mentioned compounds are solids and hence found in particulated form; the other materials mentioned are present as gases.

The most common sources of chlorine in the atmosphere are from operations in which it is manufactured or used to produce other chemicals. Also as chlorine is used in water purification plants, in sewage plants and in swimming pools, equipment failure sometimes leads to leakage of chlorine into the atmosphere.

Hydrogen chloride is evolved in numerous industrial chemical processes, but it is so easy to recover, that little reaches the atmosphere.

The main effects of chlorine and its compounds are respiratory irritation from chlorine, corrosion by hydrogen chloride and damage to vegetation from chlorine and hydrogen chloride.

Oxides of Nitrogen

It is probable that oxides of nitrogen are the second most abundant atmospheric contaminants in many cities, ranking next to sulphur dioxide. Generally the highest concentration of nitrogen oxides in gaseous emissions occurs in effluents from industries where nitric acid is produced or used in chemical reactions. The next highest concentration is in automobile exhausts. The major effluents come from large power plants, and then to a small extent those from low heat burners and furnaces.

Out of seven oxides of nitrogen (N_2O, NO, NO_2, NO_3, N_2O_3, N_2O_4, N_2O_5), only nitric oxide and nitrogen dioxide arise from many human activities and are classified as pollutants. In atmospheric analyses they are usually reported as 'total oxides of nitrogen' or "NO_x". It is a standard practice in the chemical industry to absorb and recover significant quantities of oxides of nitrogen.

Ammonia and ammonium salts are not important air contaminants. Ammonia is an important raw material in fertilizer and organic chemical industries and in the manufacture of nitric acid by the oxidation process; its recovery is a matter of fundamental important in the economical operations of such processes.

Carbon Monoxide

Carbon Monoxide, an odourless and colourless gas has its major origin in the incomplete combustion of carbonaceous materials. It is a highly poisonous gas and is generally classified as an asphyxiant.

The chief source of carbon monoxide in the atmosphere is combustion, especially due to automobile exhausts. However, except for motor vehicles and other internal combustion engines, very little carbon monoxide is found in the gaseous emissions from properly adjusted, properly operated installations. Although certain industrial operations, such as electric and blast furnaces, some petroleum refining operations, gas manufacturing plants, and coal mines, are potential contributors of carbon monoxide to the atmosphere, automobile exhausts are by far the most important source.

Ozone

The origin of ozone that is found in the air has not been clarified, but it is likely that combustion and sunlight are involved in its production. Ozone is poisonous and smelly. It exists in great abundance under natural conditions in the upper atmosphere.

Aldehydes

These are produced by the combustion of gasoline, diesel oil, fuel oil, and natural gas. Incomplete oxidation of motor fuel and lubricating oils leads to the formation of aldehydes. Lower aldehydes may be present in the atmosphere in concentrations about as high as sulphur dioxide. They may also be formed in the air because of photochemical reactions. Formaldehyde is irritating to the eyes.

Organic Vapours

These contaminants include a large number of chemical compounds, including paraffins, olefins, acetylenes, aromatic hydrocarbons, chlorinated hydrocarbons, etc. They are produced by combustion processes (particularly in automobiles), household incinerators, and petroleum processes. It is also probable that they undergo changes in the atmosphere which contribute to the formation of smog.

Radioactive Gases

A major source of radioactive gases and particulate is the nuclear power reactor and related fuel handling facilities. The other sources are experimental accelerators, testing of nuclear bombs in the atmosphere, agricultural, industrial and medical use of radioactive isotopes. Another source of radioactive particulates and gases that is increasing in importance, is the nuclear fuel reprocessing plant.

Table 7.2: Sources of Air Pollution

No.	*Aerosols*	*Gases and Vapours*
1.	Combustion processes (domestic burning, thermal power plants, cars, trucks, aeroplanes, and railways. Also refuse burning)	Dust, fume SO_2, NO_2, CO, smoke organic vapours, odours
2.	Chemical processes (paper mills, cement, fertilisers, etc.)	Dust, fume, Process dependent mist (SO_2, CO, NH_3, NO_2, organic vapour, odours)
3.	Petroleum operations	Dust, mist SO_2, H_2S, NH_3, CO, hydro-carbons, mercaptans
4.	Metallurgical processes	Dust, fume SO_2, CO, fluorides, (aluminium refineries, organic vapours steel plants)
5.	Mineral processing	Dust, fame Process dependent (SO_2, CO, fluorides, organic vapours
6.	Food and feed operations	Dust, mist, Odourous materials
7.	Agricultural activities (a) Crop spraying (b) Field burning	Organic phosphates, Dust, mist chloronated hydro Smoke, sulphur oxides, fly ash organic vapours
8.	Nuclear energy programmes (a) Fuel fabrication (b) Ore preparation (c) Bomb explosion	 Dust Fluorides Iodine-131 and Argon-41 Radioactive gases (Sr-90, Cs-137, C-14, etc.)

Primary and Secondary Air Pollutants

Air pollutants can also be broadly classified into two general groups-primary air pollutants and secondary air pollutants.

Primary air pollutants are those emitted directly from identifiable sources.

Examples of primary air pollutants

1. Finer particles (less than iGum in diameter)
2. Coarse particles (greater than 100 in diameter)
3. Sulphur compounds
4. Oxides of nitrogen
5. Carbon monoxide
6. Halogen compounds
7. Organic compounds
8. Radioactive compounds

Finer aerosols include particles of metal, carbon, tar, resin, pollen, bacteria, etc.

Secondary air pollutants are those which are produced in the air by the interaction among two or more primary pollutants, or by reaction with normal atmospheric constituents, with or without photoactivation.

Examples of secondary air pollutants:

1. Ozone
2. Formaldehyde
3. PAN (proxy acetyl nitrate)
4. Photochemical smog
5. Formation of acid mists (H_2SO_4) due to reaction of sulphur dioxide and dissolved oxygen, when water droplets are present in the atmosphere.

Smog

Smog is a synchronym of two words, smoke and fog. It can be of two types, photochemical or coal induced.

Photochemical smog is restricted to highly motorised areas in metropolitan cities, e.g., Los Angeles. It occurs under adverse meteorological conditions when the air movement is restricted. Smog is caused by the interaction of some hydrocarbons and oxidants under the influence sunlight giving rise to dangerous peroxy acetyl nitrate, (PAN). Its main constituents are nitrogen oxides, PAN, hydrocarbons, carbon monoxide and ozone. It reduces visibility, causes eye irritation, damage to vegetation and cracking of rubber.

The fog from burning coal covers urban areas at night or in cold days when the temperature is below 10°C and when calm meteorological conditions prevail, e.g., London (December 1952). This fog consists of smoke, sulphur compounds and fly ash. Prolonged exposure for smog may result in a high mortality rate especially among the elderly and those who have histories of Chronic bronchitis, asthma, bronchopneumonia and other lung or heart diseases.

Stationary and Mobile Sources

Emissions (air pollutants) may be classified by source, as stationary or mobile. Another method of classifying emission sources is by:

1. Point sources (large stationary sources)
2. Area sources (small stationary sources and mobile sources with indefinite routes)
3. Line sources (mobile sources with definite routes)

The following chart gives a picture of emission inventory sources classification.

Air pollution Due to Automobiles

This section discusses the contribution of automotive emissions in aggravating the air pollution menace. Various causes of the genesis and exodus of these pollutants have been identified and methods to control them have been found. Certain modifications rif-Hp engine design and operating variables are suggested. The possibilities of some alternatives as suitable substitutes for modern automobiles are also explored and future strategy in the Indian context has been proposed.

Emission Inventory Source Classification

Total Sources			
Stationary Sources		*Mobile Sources*	
Point sources	*Area Sources*	*Line Sources*	*Area Sources*
1. Industrial processing	1. Residential heating coal	1. Highway vehicles	1. Motor vehicles light-duty medium-duty locomotives
2. Power plants	2. Institutional & Commercial heating, coal, oil, gas	2. Railroad gas	2. Railyard locomotives
3. Fuel Combustion oil heavy-duty (industrial) coal, oil, gas	3. On site incineraiion	3. Channel vessels	3. Port Vessels
4. Solid waste disposal	4. Open burning		4. Aircraft (airports)
5. Municipal incinerators open burning	5. Evaporative losses		5. Miscellaneous
6. Miscellaneous	6. Miscellaneous		

The three main types of automotive vehicles being used in our country are (a) passenger cars powered by four stroke gasoline engines, (b) motor cycles, scooters and autorickshaws powered mostly by small two stroke gasoline engines, and (c) large buses and trucks powered mostly by four stroke diesel engines.

Emissions from gasoline powered vehicles are generally classified as: (a) Exhaust emissions, (b) Crank-case emissions and, (c) Evaporative emissions. The amount of pollutants, that an automobile emits depends on a number of factors, including the design and operation. Of the hydrocarbons emitted by a car with no controls, the exhaust gases account for roughly 65%, evaporation from the fuel tank and carburetor for roughly

15% and blow by or crankcase emission (gases that escape around the piston rings) for about 20%. Carbon monoxide, nitrogen oxides and lead compounds are emitted almost exclusively in the exhausts gases. Table 7.3 gives the emissions from a typical Indian car at different conditions.

Table 7.3: Emissions from a Typical Indian Car

Speeds Hydrocarbons (kmph)	Carbon monoxide (ppm)	Oxides of nitrogen N	(ppm)
Idle	3200	3.50	75
32	2825	1.39	475
48	2475	0.28	1375
64	2175	0.18	1600
80	2000	0.18	1945

Diesel powered vehicles create relatively minor pollution problems compared to gasoline powered ones. The diesel engines exhausts only about a tenth of the amount of carbon monoxide exhausted by a gasoline engine, although its hydrocarbon emissions may be more than those of the gasoline engine. Blowby is negligible in the diesel, since the cylinders contain only air on the compression stroke. Evaporative emissions are also low because the diesel engine uses a closed injection fuel system and because the fuel is less volatile than gasoline. The major problems of diesel engines are smoke and odour.

Air pollution from major Industrial Operations

Even though there are many different sources which contribute to air pollution, industries contribute a major share. There are a number of industries which are responsible for air pollution, but it is not possible to cover all of them. Table 7.2 shows the specific pollutants found in the emissions of some typical industrial and other sources.

Effects of air pollution on human Health

Air pollution is one of the greatest environmental evils. The air we breathe has not only life-supporting properties but also life damaging properties. Under ideal conditions the air we inhale has a qualitative balance that maintains the well-being of man. But when the balance among the air components is disturbed, or in other words, if it is polluted, it may affect human health.

An average man breathes 22,000 times a day and takes in 16 kg of air each day. It far exceeds the consumption of food and water. It has been estimated, that a man can live for five weeks without food and five days without water, but only for five minutes without air.

All the impurities in the inhaled air do not necessarily cause harm. Depending upon the chemical nature of the pollutants, some may be harmful when present in the air in small concentrations and others only if they are present in high concentrations. The duration of exposure of the body to polluted air is also an important factor. Therefore, the prime factors affecting human health are:

1. Nature of the pollutants
2. Concentration of the pollutants
3. Duration of exposure
4. State of health of the receptor
5. Age group of the receptor

Generally speaking susceptibility to the effects of air pollution is great among infants, the elderly, and the infirm. Those with chronic diseases of the lungs or heart are thought to be at great risk. Pre-school and school children appear to be both sensitive and specifically reactive to air pollution health effects. Another point to be noted is that the effect of air pollution on human health is worst during the wintér season, when pollution levels reach a climax.

An objectionable odour, visibility reduction, eye irritation or vegetation damage are useful guides to the likelihood or severity of health effect. A grey ball over a city or an industrial area can have a depressing effect and impair the enjoyment of life. There is no doubt, however, that the urgency with which steps are taken to improve air quality will depend very much on how serious the risk of ill-health from air pollution is thought to be.

Table 7.4. Air pollution Episodes

S.No.	*Month and year*	*Place*	*Mortality*
1.	December, 1930	Meuse Valley (Belgium)	63
2.	October, 1948	Donora (Pennsylvania)	20
3.	November, 1950,	Poza Rica (Mexico)	22
4.	December, 1952	London	4000
5.	November, 1953	New York	220
6.	January, 1956	London	1000
7.	December, 1957	London	750
8.	December, 1962	London	700
9.	January, 1963	New York	300
10.	November, 1966	New York	168
11.	December, 1984	Bhopal (India)	2000

Health Effects

1. Eye irritation.
2. Nose and throat irritation.
3. Irritation of the respiratory tract.
4. Gases like hydrogen sulphide, ammonia and mercaptans cause odour nuisance even at low concentrations.
5. Increase in mortality rate and morbidity rate.
6. A variety of particulates particularly pollens, initiate asthmatic attacks.
7. Chronic pulmonary diseases like bronchitis and asthma, are aggravated by a high concentration of

SO_2, NO_2, particulate matter and photochemical smog.

8. Carbon monoxide combines with the haemoglobin in the blood and consequently increases stress on those suffering from cardiovascular and pulmonary diseases.
9. Hydrogen fluoride causes diseases of the bone (fluorosis), and moi aisg of teeth.
10. Carcinogenic agents cause cancer.
11. Dust Particles cause respiratory diseases. Diseases like silicosis, asbestosis, etc., result from specific dusts.
12. Certain heavy metals like lead may enter the body through the lungs and cause poisoning.

Health Effects Of Specific Pollutants

Sulphur Dioxide

Sulphur dioxide is an irritant gas which affects the mucous membranes when inhaled. Under certain conditions, some of the air borne sulphur dioxide gas is oxidized to sulphur trioxide. Each of these two gases in the presence of water vapour or water, forms sulphurous and sulphuric acid respectively. Sulphur trioxide is a very strong irritant, much stronger than sulphur dioxide, causing severe bronchospasms at relatively low levels of concentration.

Carbon Monoxide

Carbon monoxide has a strong affinity for combining with the haemoglobin of the blood to form carboxyhaemoglobin, COHb. This reduces the ability of the haemoglobin to carry oxygen to the body tissues. CO has about two hundred times the affinity of oxygen for attaching itself to the haemoglobin, so that low levels of CO can still result in high levels of COHb. Carbon monoxide also effects the central nervous system. It is also responsible for heart attacks and a high mortality rate.

Oxides of Nitrogen

Of the seven oxides of nitrogen known to exist in the ambient air, only two are thought to affect human health. These are nitric oxide (NO) and nitrogen dioxide(NO_2). While some questions remain about haemoglobin reactions with oxides of nitrogen, there is no positive evidence that nitric oxide exposure is a health hazard associated with community air pollution.

Nitrogen dioxide is known to cause occupational disease. Among occupations with NO, hazards are the manufacture of nitric acid, exposures of farmers to silage that has high nitrate fertilisation, electric welding, and mining utilizing nitrogen compounds as explosives. It is estimated that eye and nasal irritation will be observed after exposure to about 15 ppm of nitrogen dioxide and pulmonary discomfort after brief exposures to 25 ppm of nitrogen dioxide.

Hydrogen Sulphide and Mercaptans

Hydrogen sulphide is a foul smelling gas. It is well known for its rotten egg like odour. Exposures to hydrogen sulphide for short periods can result in fatigue of the sense of smell.

Other sulphur compounds that are of interest in air pollution mainly because of their strong odours are methyl mercaptan (CH, SH) and ethyl mercaptan (C_2H_5SH). But it has been reported that at the concentrations at which they are odour nuisances, they have no other effect on human health. In fact, mercaptans are often added to natural or manufactured gas supplies, so that leakage of gas will be noticed.

Ozone

Ozone is a gas that has an irritant action in the respiratory tract, reaching much deeper into the lungs than the oxides of sulphur.

Fluorides

Fluorides present in air, range from these which are extremely irritant and corrosive like hydrogen fluoride to relatively non-reactive compounds. But, fluorine is a cumulative poison even under condition of prolonged exposure and in sub-acute concentrations.

Lead

The main source of lead in urban atmosphere is the automobile. It creates urban concentrations of inorganic lead of about 1-3mg/m with high values in area of heavy traffic. Inorganic lead acts as an agent which causes a variety of human health disorders. The effects include gastrointestinal damage, liver and kidney damage, abnormalities in fertility and pregnancy, and mental development of children gets affected.

Hydrocarbon Vapours

Some of the hydrocarbon vapours in the atmosphere have health implications. The effect of formaldehyde is primarily irritating. It is a major contributor to eye and respiratory irritation caused by Photochemical smog.

Carcinogenic Agents

Carcinogenic agents are responsible for cancer. For example, the poly-cyclic organic compound, 3, 4-benzpyrene. The origin of these compounds is incomplete combustion of hydrocarbons and other carbonaceous materials. They are also reported to be present in exhaust discharges from I.C. engines. In addition to poly-cyclin organic compounds, it has been found that some aliphatic hydrocarbons are also carcinogenic.

Insecticides

Insecticides are not only harmful for insects but also poisonous for man, e.g., DDT (Dichloro diphenyl trichloro-ethane). They can affect the central nervous system and may attack other vital organs. In fact, DDT has been found in mother's milk in western countries and even in our own country. Hence, the use of DDT has been banned in the USA. The United Nations Environment Programme (UNEP) has warned against the undesirable effects of indoor spraying of DDT. it has also been reported that indoor spraying affects domestic livestock.

According to a study conducted at the Industrial Toxicology Research Centre, Lucknow (India), the accumulation of pesticides in the environment due to their growing use for agricultural purposes can also cause premature labour and abortion, due to high concentration of pesticides in the body of expectant mothers.

Radioactive Isotopes

The important radioactive isotopes that may reach ambient air are Iodine 131, phosphorous 32, Cobalt 60, Strontium 90, radium 226, Carbon 14, Sulphur 35, Calcium 45 and Uranium. The major sources of radioactive air pollutants are:

(a) Nuclear reactors

(b) Experimental accelerators

(c) Scientific and medical use of radioactive isotopes

(d) Agricultural and industrial use of radioactive isotopes as tracers, and

(e) Testing of nuclear bombs in the atmosphere.

The serious health effects are anaemia, Leukaemia and cancer. Radioactive isotopes also cause genetic defects and sterility, as well as embryo defects and congenital malformations. It also shortens the life span of an individual.

Effects of Air Pollution on Animals

Interest in the effects of air pollution on animals has generally developed as a corollary to the concern about its influence on human health. Most of the information concerning the natural exposure of animals to air pollution is contained in the reports of some major air pollution disasters e.g., Donora, London and Poza Rica. Recently, considerable information has been reported from medical research laboratories which describes the results of experimental exposure of small animals to various air pollutants. Animals used for laboratory work were mice, rabbits, rats, guinea pigs and monkeys.

Air Pollution Effects on Farm Animals

The process by which farm animals get poisoned is entirely different from that by which human beings exposed to polluted atmospheres are poisoned. In case of farm animals it is a two-step process.

1. Accumulation of the air-borne contaminant in the vegetation and forage.
2. Subsequent poisoning of the animals when they eat the contaminated vegetation.

In case of human beings working in polluted atmospheres in factories, the concern is for the harmful substances that are directly inhaled, whereas in the case of 'the farm animals, the danger obviously is not in inhaling the polluted air, but rather than ingestion of forage which has been contaminated with pollutants like fluorine from the air.

The three pollutants responsible for most livestock damage are fluorine, arsenic and lead. These pollutants originate from industrial sources or from dusting and spraying.

Fluorine

Of all farm animals, cattle and sheep are the most susceptible to fluorine toxicosis. Horses appear to be quite resistant to fluorine poisoning. Poultry are probably the most resistant to fluorine, of all farm animals.

Symptoms of Acute Fluorine Poisoning

Lack of appetite, rapid loss in weight, decline in health and vigour, lameness, periodic diarrhoea, muscular weakness and death, characterize the acute form of fluorine poisoning. It may also result in considerable increase of bone fluorine. But acute poisoning due to fluorides is unlikely in a majority of the cases.

Symptoms of Chronic Fluorine Poisoning

Fluorine is a cumulative poison under conditions of continuous exposure to subacute doses. Fluorine is also a protoplasmic poison. It has a marked affinity for calcium and interferes with normal calcification. Animals have been reported to be more iesibiaju UMH humans to dental mottling. Cattle and sheep are the most frequently affected animals. Teeth in the process of formation are easily affected. Hence tooth symptoms are a sensitive and unique criterion of chronic fluorosis. Excessive wearing of incisor and molar teeth may occur at high levels of fluorine intake.

Bone lesions may develop at any age. A bony overgrowth may be observed on the leg bones, jaw bones and ribs. Their appearance indicates that there has been a high fluoride intake over a long period. Lameness may occur as a result of this overgrowth on the leg bones. This may be followed by stiffness of joints. Effects of fluorides on skeletal structures are usually of a permanent nature.

Symptoms of advanced fluorosis include lack of appetite, general ill-health due to malnutrition, lowered fertility, reduced milk production and growth retardation. The symptoms developed in other species such as rabbits, horses and poultry are similar to those in cattle and sheep, i.e., mottling, staining and wearing of teeth, bony overgrowths on the skeleton, stiffness and lethargy and general ill-health from malnutrition and starvation.

While diagnosis of chronic fluorine poisoning is done, one has to be careful. This is because drinking water supplies may contain unduly high levels of fluorides. Also, mineral supplements may contribute appreciable amounts of fluorine to the diet. Hence, a check of all possible sources must be made when fluorosis.

Arsenic

Arsenic occurs as an impurity in many ores and in coal. It has been reported to cause poisoning of livestock near various industrial processes and smelter. Just like most industrial air contaminants, arsenic may spread over a considerable area from a stack source. Arsenic in dusts or sprays on plants can lead to poisoning of cattle.

Symptoms of Acute Arsenic Poisoning

In acute cases the symptoms are severe salivation, thirst, vomiting, uneasiness, feeble and irregular pulse and respiration. The animal stoops, lies down and gets up. There is diarrhoea, and the faeces have a garlic odour and are sometimes bloody. The ears become cold, the body trembles and develops abnormal temperature and convulsion. Death may occur in few hours or days.

Symptoms of Chronic Arsenic Poisoning

Arsenic appears to have a depressing effect upon the central nervous systems. The animal becomes dull, and exhibits a lack of appetite, with a resulting weight loss. Also, there may be chronic cough and diarrhoea, may occur continuously. There may be thickening of the skin, anaemia and abortion or sterlity. Chronic poisoning can result in eventual paralysis and death.

Tolerance limit

Arsenic as well as its soluble compounds are extremely poisonous. It has been reported that sheep have been poisoned by as little as 0.25-0.50 g of arsenic daily, whereas cattle and horses may tolerate 1.3-1.9 g daily. It has also been reported from laboratory experiments, that 10 mg of arsenic per kilogram of body weight per day has produced chronic arsenic poisoning in guinea pigs. A detailed investigation of the effects of arsenic poisoning on farm animals is necessary to establish standard tolerance limits.

Lead

Lead contamination of the atmosphere takes place on account of various industrial sources such as smelters, coke ovens and other coal combustion processes. Lead is also used in dust and sprays containing lead arsenate.

Symptoms of Acute Lead Poisoning

In case of acute lead poisoning the onset is sudden and the course relatively short. Prostration, staggering and inability to rise are prominent symptoms. The pulse is always fast but weak. Some animals may fall suddenly, stiffen the legs and have convulsions. There is complete loss of appetite, paralysis of the digestive treat and diarrhoea. Other nervous symptoms in cattle are grinding of the teeth and rapid chewing of the cud.

In case of horses, it may result in complete loss of appetite, nervous depression, lethargy and death.

Symptoms of Chronic Lead Poisoning

Chronic lead poisoning has been observed frequently in horses that have been grazing on forage near smelters, lead mines, and in orchards that have been sprayed. Paralysis of the muscles of the larynx and difficulty in breathing are the main symptoms. Convulsions may occur from the paralysis of the throat, and the difficulty in breathing may be unusually severe and persistent, during and after exercise.

Tolerance Limits

Lead is a cumulative poison. Therefore, the continuous ingestion of even very small daily doses will be finally as effective as one toxic dose. Hence in case of slight contamination death can occur after many months and in case of severe contamination death can occur within 24 hours.

It has also been reported that inadequate calcium in diet increases the retention of lead in the animal body. Low-calcium diets may reuse in lead storage of as much as five times as compared to that found in animals that receive sufficient amounts of calcium. However, extra quantities of calcium above the amount considered sufficient, will not offer additional protection against poisoning.

Effects of Air Pollution on Plants

Air pollution has been known to have an adverse effect on plants. At first, it was only sulphur dioxide that was considered a dangerous pollutant. Now, with the advent of various pesticides and new industrial processes, the range of harmful pollutants has multiplied tremendously. Sometimes, vegetation over 150 km away from the source of the pollutant has been found to be affected.

Industrial pollution, particularly from smelters, has caused complete destruction of vegetation in some cases e.g., at Ducktown, Tennessee. The international dispute between Canada and the United States over the damage caused by the Trail, British Columbia Copper Smelter is well known. Los

Angeles smog has caused widespread damage to some crops and forests in Southern California. In fact, many books and reports have been published in USA on air pollution injury to vegetation. In some cases, economics of the damage caused to crops and plants has been estimated. Compensation amounting to thousands of dollars has been paid for damage done to the crops. In our own country, there are many reports of the effect of pollutants like cement dust on plants.

Effect of Environment on Plants

The primary factor which controls gas absorption by the leaves is the degree of opening of the stomata. When the stomata are wide open, absorption is maximum and vice versa. Consequently, the same conditions that enhance the absorption of the gas (CO_2 for photosynthesis), predispose the plant to injury (by absorbing a pollutant gas like SO_2). The conditions that cause the stomata to open are high light intensity (especially in the morning hours), high relative humidity, adequate moisture supply to the roots of the plant and moderate temperatures.

Most plants close their stomata at night and are therefore much more resistant at night than in the day time. But some plants like the potato, which do not close their stomata at night are as sensitive in the dark as in the light.

Air Pollutants Affecting Plants

Sulphur dioxide	Fluoride compounds (like hydrogen fluoride)
Ozone	Chlorine
Hydrogen chloride	Nitrogen oxides (NO, NO_2, etc.)
Ammonia	Hydrogen sulphide
Hydrogen cyanide	Mercury
Ethylene	PAN (peroxy acetyl nitrate)
Herbicides (sprays of weed killers)	Smog

The above pollutants interfere with plant growth and the phenomenon of photosynthesis. Smog, dust, etc., reduce the amount of light reaching the leaf and also by clogging the stomata may reduce carbon dioxide intake to some extent and thus interfere with photosynthesis.

Forms of Damage to Leaves

Damage to leaves takes several forms

1.	Necrosis	Necrosis is the killing or collapse of tissue.
2.	Chlorosis	Chlorosis is the loss or reduction of the green plant pigment, chlorophyll. The loss of chlorophyll usually results in a pale green or yellow pattern. Chlorosis generally indicates a deficiency of some nutrient required by the plant. In many respects, it is analogous to anaemia in animals.
3.	Abscission	Leaf abscission is dropping of leaves.
4.	Epinasty	Leaf epinasty is a downward curvature of the leaf due to higher rate of growth on the upper surface.

Tissue severely injured by air pollutants often has a characteristic colour. Bleaching is associated with sulphur dioxide, yellowing with ammonia, browning with flouride. A silvering or bronzing of the under surface of some leaves is associated with PAN injury. Phytotoxicant is the name given to plant damaging substances.

Kinds of Injury to Plants

Acute Injury

It results from short-time exposure to relatively high concentrations, such as it might occur under fumigation conditions. The effects are noted within a few hours to a few days and may result in visible markings on the leaves due to a collapse and death of cells. This leads to necrotic patterns i.e., areas of dead tissue.

Chronic Injury

It results from long-term low level exposure and usually causes chlorosis or leaf abscission.

Growth or Yield Retardation

Here the injury is in the form of an effect on growth without visible markings (invisible injury). Usually a suppression of growth or yield occurs.

While deciding about the type of injury the plant has suffered, one must carefully distinguish between adverse effects caused by air pollution and adverse effects resulting from other factors. Plant diseases may show symptoms which are very similar to those caused by air pollution. High temperatures, poor plant care, shortage of nutrients and water may also cause an appearance which is similar to that of a plant damaged by air pollution. It may also be due to insect damage. Therefore, while diagnosing the effects of air pollution one must consider factors such as plant disease, nutrition history, weather damage, insect damage, and the nature of pollutants in the area.

Effects of Air Pollutants on Plants

The effects of various pollutants like sulphur dioxide, ozone, fluorides, etc., on plants is shown in Table 7.5.

Table 7.5: Effects of Pollutants on Plants

	Pollutant	*Dose*	*Effect*
1.	Sulphur dioxide	Mild	Interveinal chlorotic blanching of leaves, Severe Necrosis in interveinal areas and skeletonized leaves
2.	Ozone	Mild	Flecks on upper surfaces, premature aging and suppressed growth Sei-ere Collapse of leaf necrosis and bleaching
3.	Fluorides	Cumulative	Necrosis at leaf tip effect

4.	Nitrogen dioxide	Mild	Suppressed growth, leaf bleaching
5.	Ethylene	Mild	Epinasty, leaf abscission
6.	PAN	Mild	Bronzing of lower leaf surface (upper surface normal), suppressed growth. Young leaves more susceptible

Effect on Sulphur Dioxide

Sulphur dioxide produces two types of injury on the leaves of plants, acute and chronic, depending on the concentration and period of explore. The acute injury is characterized by the killing of marginal or interveinal areas of the leaf. Immediately after fumigation these areas will get a dull, water-soaked appearance, subsequently they dry up and usually bleach to an ivory colour, though some species finally assume a brown or reddish-brown colour. Chronic injury is caused by the slow, long-continued absorption of sub-lethal amounts of gas or by absorption of an amount of gas somewhat less than that necessary to cause acute injury.

Sulphur dioxide is phytotoxic in concentrations above 0.1-0.2 ppm. Below about 0.4 ppm. it tends to be oxidised in the cells as rapidly as it is absorbed, but interferences with functions such as photosynthesis is slight. Chronic injury, if any, is generally exhibited with these small concentrations.

Above about 0.4 ppm, acute injury occurs more frequently. Sometimes. temporary interference with photosynthesis or 'invisible injury' can occur. However, the mechanism by which sulphur dioxide affects the plants has not yet been well understood, though various explanations have been put forth by many investigators.

Effect of Hydrogen Fluoride

Hydrogen fluoride behaves somewhat similar to sulphur dioxide except that with a few species of plants, it is effective in causing lesions and interfering with photosynthesis in concentrations two or three orders less than in the case of sulphur dioxide. With most species it is up to 10 times as effective as sulphur dioxide. However, recovery of plants from the fluoride effect is much slower than from sulphur dioxide. This difference in rate of recovery is probably explained by the fact that sulphite in the leaves is rapidly oxidised to the relatively non-toxic sulphate, whereas fluorides can be removed only by the slower process of valorization by some obscure chemical reaction. Forage may be rendered unsafe for animal feeding if more than 50 ppm of fluorine is absorbed.

Effect of Ozone

Ozone is phytotoxic in exposures of a few hours at about 0.2 ppm. Injury due to ozone is quite different from typical smog injury. The lesions are generally confined to the upper surface. The effects of ozone on plants in mild and severe doses is indicated in Table 7.5.

Economic Effects of Air Pollution

Air pollution damage to property is very important economic aspect of pollution. In the United States of America, Great Britain, France and many other European countries, this problem has been investigated in detail and successful attempts have been made to translate observable air pollution damage into terms of economic impact. Air pollution damage to property covers a wide range corrosion of metals, soiling and eroding of building surfaces, fading of dyed materials, rubber cracking, spoiling or destruction of vegetation, effects on animals, as well as interference with erodiation and services. Another important economic effect of air pollution is deterioration of works of art. In our country today there is an urgent need to investigate and study this problem in detail and express the damage to property in economic terms, as very little work has been done in this direction so far.

Mechanisms of Deterioration in Polluted Atmospheres

Air pollutants cause damage to materials by five mechanisms:

1. *Abrasion:* Solid particles of sufficient size and travelling at high velocities can cause abrasive action.

Also, large sharp edged particles embedded in fabric can accelarate wear.

2. *Deposition and removal:* Solid and liquid particles deposited on a surface may not damage the material itself but it may spoil its appearance. However, the removal of these particles may cause some deterioration. Although a single washing or cleaning may not cause noticeable deterioration, frequent cleaning ultimately does.

3. *Direct chemical attack:* Some air pollutants react directly and irreversibly with materials to cause deterioration. For example, the bleaching of marble by sulphur dioxide, the tarnishing of silver by hydrogen sulphide and the etching of a metallic surface by an acid.

4. *Indirect Chemical attack:* Certain materials absorb some pollutants and get damaged when the pollutants undergo chemical changes. For example, sulphur dioxide absorbed by leather is converted to sulphuric acid, which deteriorates the leather.

5. *Corrosion:* the atmospheric deterioration of ferrous metals is by an electrochemical process, called corrosion. This due to the action of air pollutants facilitated by the presence of moisture.

Table 7.6 presents a generalized picture of air pollution damage to various materials.

Table 7.6: Air Pollution Damage to Various Materials

Materials	*Principal air pollutants*	*Effects*
Metals	SO_2, acid gases	Corrosion, spoilage of surface, loss of metal, tarnishing
Building materials	SO_2, acid gases	Discolouration, leaching
Paint	SO_2, H_2S	Discolouration particulates

Textiles and dyes	SO_2 acid gases and NO_2, Ozone	Deterioration, Textile reduced tensile strength, and fading
Rubber	Oxidants, ozone	Cracking, weakening
Leather	SO_2, acid gases	Disintegration powdered surface
Paper	SO_2, acid gases	Embrittlement
Ceramics	Acid gases	Change in surface appearance

Factors Influencing Atmospheric Deterioration

There are four prime factors which influence the rate of attack of damaging pollutants. They are moisture, temperature, sunlight, and air movement.

1. *Moisture:* The presence of moisture in the atmosphere greatly helps the process of corrosion. Without moisture in the atmosphere, there would be little atmospheric corrosion, if any, even in the most severely polluted environments. In case of sulphur dioxide and various combinations of particulates, investigations have revealed that the rate of corrosion of metals will increase as relative humidity in the air increases.

2. *Temperature:* Temperature affects the rate of the chemical reaction and consequently it affects the rate of deterioration.

3. *Sunlight:* In addition to producing damaging agents such as ozone through a series of complex photo-chemical reactions in the atmosphere, sunlight can cause direct deterioration of certain materials.

   But in cases of fading of certain dyes or rubber cracking, damage due to direct sunlight cannot always be distinguished from that caused by ozone.

4. *Air movement:* Wind direction is an important factor to be considered in places where deterioration is caused by pollutants released from nearby factories. For example, damage to agricultural crops in the vicinity of factories. Similarly, wind speed is also an important factor in determining the impact of air pollutants on the receiving surfaces. Pollutants present in wind blowing at high speeds will have more abrasive effects and they may also be carried over long distances. Where leather bound books are stored, air movement is a critical factor when the air is contaminated by sulphur dioxide.

Effects on Art Treasures

Air pollutants have caused great damage to art objects throughout the industrialized world. Some excellent monuments of antiquity have been silently eaten away by pollutants. This can be clearly seen in many works of art sculptures and carvings on historic buildings and cathedrals. A few examples of such works of art which have been affected by air pollution both in India and abroad have been described below.

Effect on Art Treasures in India

(a) *Taj Mahal:* The Taj Mahal a miracle marble and a historical monument is facing grave danger from pollution created by existing foundries, power house, railways yards and other industrial units. According to some environmentalists, the problem now seems to have been aggravated because of the commissioning of the Mathura Oil Refinery, about 40 km from the priceless monument. Whether the monument is going to be affected by the pollutants emitted from the refinery or not has created a controversy which is now being debated. Many reports on this have appeared in noted newspapers and journals. However, it is to be hoped that certain anti-pollution safeguards built into the refinery will minimise the effect of pollutants on the monument.

According to some reports, pollutants from existing foundries power houses and railway yards are already affecting the Taj Mahal. Black and brown patches are disfiguring the shiny white surface of the Taj. Reports in newspapers also speak of the formation of gypsum, scaling and flaking effects.

Sulphur dioxide from the various sources mentioned, mixes with the atmospheric moisture and gets converted into sulphuric acid which settles down on the exterior of the Taj Mahal. The acid reacts with the marble (calcium carbonated) and corrosion and discolouration of the monument is the ultimate result.

It looks as though more money will have to be spent on protecting the Taj Mahal from the effects of pollution than it took Shah Jehan to construct it. Perhaps the immediate solution to save the Taj Mahal lies in shifting the foundries and power houses from Agra, and in effectively supervising antipollution measures adopted by the refinery. Further, the coal locomotives being used by the railways in that region can be replaced by diesel engines. In addition to these steps, a dense green belt of suitable tree species should be created around the Taj Mahal to absorb atmospheric pollutants as trees can act as effective air scrubbers.

(b) *Belur temple:* The renowned temple of Sri Channakeshava at Belur (Hassan district, Karnataka state) is threatened with a similar hazard. A plywood factory located close to the temple emits sootladen fumes which get deposited on the sculptures in the temple and discolour the surface, inside and outside. This pollution effect is under investigation.

Effect on Art Treasures in Other Countries

The Parthenon in Athens, has been affected in recent years because of air pollution, Cleopatra's needle the large stone obelisk moved from Alexandria, Egypt, to London, has suffered more deterioration in the damp, smoky, acid atmosphere of London in the last 90 years than in the earlier 3000 or more years of its history. Another ancient Egyptian obelisk in New York City, shown similar decay.

Similar cases have been reported from other parts of the world. The colosseum in Rome, and the San Marco Basilica in Venice, Italy, show signs of decay, possibly-due to air pollution, In France, it is reported that environmentalists have removed statues from the exterior of cathedrals and replaced them with copies. Similarly ancient temples and buildings in industrial areas of Japan are facing the hazards of atmospheric pollution. Painting, books fabrics, antique costumes and other art objects have suffered damage.

Of course, it is impossible to make adequate substitution or compensation for the losses suffered by works of art. They are irreversible losses. It is impossible to estimate the losses in economic terms for these intangible assets. In some cases it is possible to preserve them, by applying special preservatives. But they are not always successful, Sensitive art objects displayed inside buildings can be placed in hermetically sealed containers. Air conditioning can also be used as a protective measure.

Pic. 7.1 Careful composition was the prize-winning secret

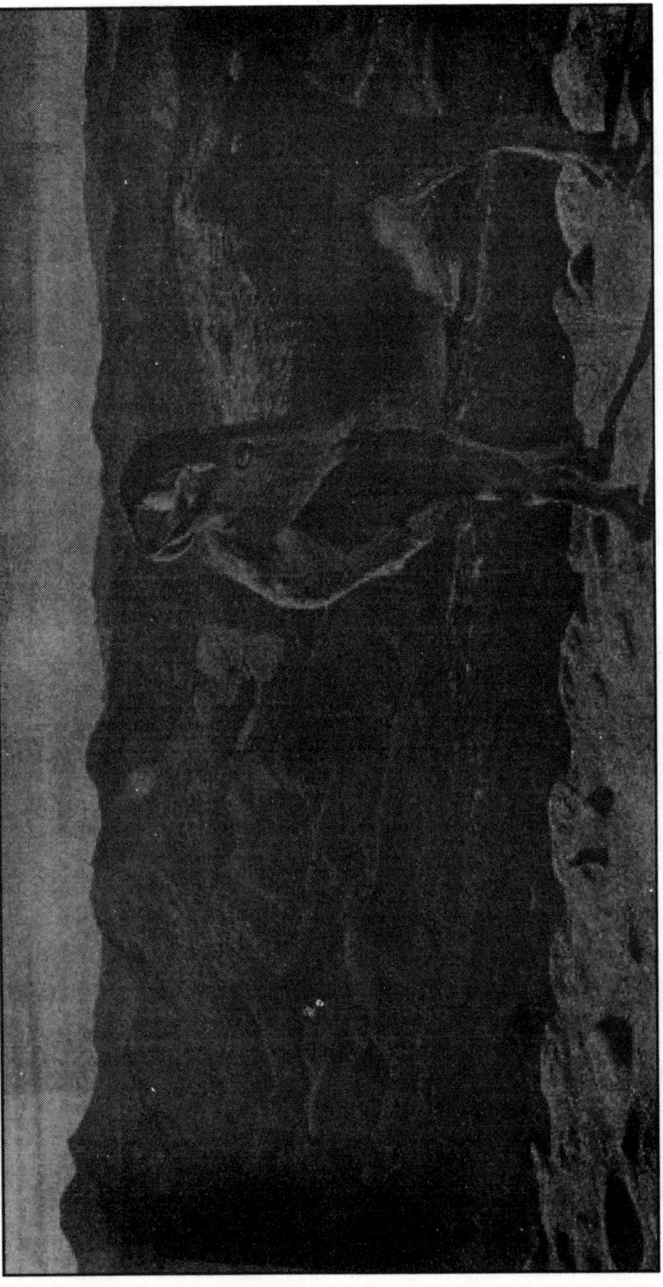

Pic. 7.2 Endangered and keystone species, like the Nilgiri tahr, constitute the basis for bio-diversity conservation

Pic. 7.3 H.V. Praveen Kumar's "The Royal Fight".

8

WATER POLLUTION

Water pollution is a phenomenon that is characterized by the deterioration of the quality of land water (rivers, lakes, marshes and ground water) or sea water as a result of various human activities. Here, we will first classify various human activities that cause water pollution, discuss water pollution according to their characteristics, and then take a general view of water pollution which is currently in progress in our modern society. Human activities related with water pollution comprise various industries such as mining, agriculture, stockbreeding, fisheries, forestry, urban human activities, manufacturing industry, construction works and various tertiary industries. Then, in the last section, we will classify waste waters according to their physical and chemical properties. Water pollution is any physical or chemical change in water that can adversely affect organisms. It is a global problem, affecting both the industrialized and the developing nations. The water pollution problems in the rich and the poor nations, however, are quite different in many respects. Heat, toxic-metals, acids, sediment, animal and human wastes, and synthetic organic compounds foul the waterways of developed nations. Human and animal wastes, sediment, and pathogenic organisms head the list in the non-industrialized nations. In these countries, unsanitary water and malnutrition account for most of the illness and death. Water has many different uses, each requiring different levels of purity. For example, water from the Ohio River may be clean enough to wash steel but may be toxic to fish and wildlife: water suitable for boating or fishing may be unsuitable for swimming : water safe to swim in, may be too polluted to drink.

Like air pollutants, water pollutants come from numerous natural and anthropogenic sources. Likewise, water pollutants produced in one may flow into others, creating complex international control problems that may take decades to solve.

Major Water Pollutants

Following are the eight major types of water pollutants:

1. Oxygen-demanding wastes (domestic sewage, animal manure, and some industrial wastes)
2. Disease-causing agents (bacteria, parasites, and viruses)
3. Inorganic chemicals and minerals (acids, salts, and toxic metals)
4. Organic chemicals (pesticides, plastics, detergents, industrial wastes, and oil)
5. Plant nutrients (nitrates and phosphates)
6. Sediments (soils, silt, and other solids from land erosion)
7. Radioactive substance
8. Heat (from industrial and power plant cooling water)

Although sedimentation is the major water pollutant in terms of quantity, other water pollutants can pose serious threats to plants and animals (including humans) when present in minute quantities. Some organisms have low tolerance levels to chemicals such as oil, some pesticides, and lead and mercury compounds. Other organisms with higher tolerance levels can be affected when the levels of some pollutants are biologically concentrated in food chains.

Water pollutants can be classified as degradable (either rapidly or slowly), or nondegradable. Rapidly degradable (non-persistent) pollutants can be broken down fairly quickly by natural chemical cycling processes as long as the pollutants do not overload the system. They include domestic sewage, other oxygen-demanding wastes, plant nutrients, and some synthetic organic chemicals. To control these pollutants we need to prevent overload. Normally these pollutants are degraded into harmless or less harmful forms, but sometimes they are converted into more harmful forms. For example, microorganisms in acidic waters can convert moderately harmful mercury metal and inorganic mercury compounds into very toxic organic methyl mercury.

Slowly degradable (persistent) pollutants remain for long periods of time but are eventually broken down or reduced to harmless levels by natural processes. Examples include some radioisotopes and many synthetic organic chemicals, such as DOT, PCBs, phenols, and oldertype detergents. Some of these substances, such as DOT, may be magnified up the food chain to levels that are harmful for wildlife and possibly for humans because most of these pollutants are extremely difficult to remove by waste treatment methods, control measures must emphasize (1) preventing or minimizing their entry into the environment, (2) long-term storage until levels are safe, and (3) research to determine which of these pollutants are dangerous and capable of being minified up the food chain. Nondegradable pollutants are not broken down by natural purifying processes. They include some metals (such as mercury, lead, and arsenic), some salts of metal sediments, some synthetic organic compounds (such as plastics), and some bacteria and viruses. These pollutants must be controlled either by removing them through waste treatment or by preventing them from entering the environment.

Sources of Water Pollution

Water pollution can result from natural runoff; dissolved chemicals in water that percolates through the soil; and human sources, such as agriculture, mining, construction, industry, homes, and businesses. It is useful to classify pollution sources as either point sources or non-point sources. Point sources involve discharge or wastes from identifiable points. They include (1) sewage treatment plants (which remove some but not all pollutants); (2) storm water runoff from combined storm and sanitary sewer lines in urban areas; (3) industrial plants; and (4) animal feedlots where large numbers (sometimes 10,000 or more) of animals are scientifically fed in tightly restricted quarters before slaughter. Non-point sources involve the diffuse discharge of wastes from land runoff, atmospheric washout, and sources that are difficult to identify its control. They include (1) runoff of sediment from natural and human caused forest fires, construction, logging, and farming; (2) runoff of chemical fertilizers, pesticides, and saline irrigation water from croplands; (3) urban storm water runoff; (4) drainage of acids, minerals, and sediments from active and abandoned mines; and (5) spills of oil and other hazardous materials. Non-point source pollution is not recognized as a major problem, since the sources are widely spread out, difficult to identify, and hard to control.

Water Pollution by Agriculture

Agriculture has been a victim of water pollution in many instances, but sometimes, it is also responsible for polluting water. Water pollution caused by agriculture is mainly an outcome of fertilizers and agricultural chemicals such as insecticides and herbicides.

(1) Water Pollution by Fertilizers

Fertilizers given to crops are not always fully consumed, and part of it remains in the soil by being absorbed by soil colloid, and influence the quality of underground, river and sea waters when it is dissolved. Chemical fertilizers mainly consists of relatively simple compounds of nitrogen, phosphorus and potassium which are the nutritive elements of plants. Their runoff coefficients vary depending on the solubility of the fertilizer itself, the rate of absorption by plants, rate of absorption by the soil and rate of decomposition. It is generally believed that the runoff coefficient of nitrogen is 5-30%, and that of phosphorus is 0.5-5%. Especially, nitrogen having a high runoff coefficient tends to cause the eutrophication of lakes and enclosed sea areas.

Fertilizers as the sources of water pollution are called "non-point sources" in contrast to "point sources" such as sewage and industrial waste waters. The non-point sources are not only hard to eliminate, but also hard to measure their quantity. Organic fertilizers such as compost, farmyard manure, human waste and fish meal produce nitrogen and phosphorus when they are decomposed, and also become the sources of pollution. These organic fertilizers do not contain, in general, any substances.

(2) Water Pollution by Agricultural Chemicals

Agricultural chemicals are considered as a source of peculiar pollution which cannot be overlooked. Agricultural chemicals comprise a variety of chemicals such as insecticides, sterilizers and weed killers, and have played important roles in the operation of modern agriculture, but they have also caused serious environmental pollution.

The characteristics of agricultural chemicals as water-polluting substances lie in that (1) almost all of them are special compounds not occurring at all in nature, and many of the behaviours and changes of such chemicals at the time when they are discharged into the natural environment are known; (2) many of their influences on the human body because of their inherent characteristic; (3) when they are ingested into the human body, they tend to accumulate or concentrate in it; (4) they are available in a variety of kinds, and new products are developed constantly; (5) some of them are extremely stable as chemicals, and remain in nature or in the human body for a long time thereby eventually afflicting the human body; (6) it is common to all such chemicals to contain additives besides the main constituents, and the additives can contribute to affect the environment though they do not have a strong toxicity; (7) many of them require a special method of analysis.

Of the agricultural chemicals developed in the past, many of them are prohibited from being used because of their violent toxicity. Nevertheless, those remaining in the environment are still factors that contribute to environmental pollution.

In addition to above-mentioned pollutions, there are instances of water pollution caused by indirect effect of agricultural activities. For example, the change in the condition of the ground surface by reclamation of waste land, deforestation, supply of the soil to low land, and the alteration of the natural water system for irrigation purposes can be considered indirect causes of water pollution.

Water Pollution by Mining Industry

In many countries, water pollution related to the mining industry can be traced back to olden times compared with those caused by other industries.

The characteristics of waste water from mining industries varies greatly depending on the kind of mine. For instance,

(i) Metal Mining Industry

Many kinds of waste waters are discharged. For example, mine water, ore dressing water, waste water from smelters, etc. In general, it is strongly acidic (low pH), and contains metal irons, sulphuric acid and suspended solids of high concentrations.

(ii) Coal Mining Industry

Waste water from coal mines mainly consists of mine water, coal dressing water, exudative water from coal-waste heap etc., and contains a large, quantity of suspended solids. In general, the scales of operation of coal mines is much larger than that of other mining industries, and thus the coal mining industry exerts a large influence upon the natural environment.

(iii) Sulphur Mine

In general, waste water from sulphur mines contains large quantities of sulphuric acid and iron and has a low pH value. Almost all sulphur mines are deserted; so the administration of deserted mines becomes a social and environmental problem.

(iv) Petroleum and natural Gas Drilling Industries

Different from other mining industries, these industries are very often located in plains. Waste water generally consists of welled-up mine water and waste water incidental to the drilling operations. It contains large quantities of salt and organic matters such as mineral oil.

(v) Stone-quarrying and Clay-supply Industries

These industries supply lime stones, building stones, porcelain clay etc. Waste water from the sites of these industries generally contains large quantities of suspended solids.

The characteristics of industrial pollution common to the above mentioned mining industries can be summarized as follows:

(i) In many instances, the development of a mine itself changes its natural environment artificially, and such a change first affects the quality of water of the environment even if any kind of waste water is not discharged from mining activities.

(ii) In areas where mines can be developed, very often, even before the development of the mine, the quality of river water or underground water is different from that of other areas. For instance, river water or underground water in such areas contain special metals of high concentrations and is strongly acid. Thus, it is rather difficult to measure the direct influence of mining activities on the quality of water. And also, geological features and the ways of land use are usually very peculiar in mining areas. These conditions should be taken into consideration prior to surveying the quality of water.

(iii) Mining activities generate large quantities of solid wastes in most cases and such wastes from large heaps. Such wastes not only pollutes water by exudation of toxic substances, but also, the rapid generation of suspended solids in water when the heap is washed away by a heavy rainfall or a flood.

Water Pollution by Stockbreeding and Fisheries

1. When cattle or horses are fed on a big pasture, there is little possibility of the livestock causing water pollution because their excrete are spread widely all over the pasture. However, when livestock is fed intensively in a limited area it causes a number of problems. In the case of the Stockbreeding industry, the excrete and leftover feed on livestock are the sources of water pollution. Water pollution caused by Stockbreeding has the following characteristics :

(i) Degree of pollution varies largely depending on the location of the Stockbreeding farm (whether in suburban or country area), system and scale of operation and the level of breeding technique.

(ii) Composition and quantity of excrete of livestock varies depending on the type, quantity and methods of feeding and watering.

(iii) Besides water pollution, the Stockbreeding industry also generates an offensive odor, flies, noise and so on.

2. Besides Stockbreeding farms, waste water from slaughterhouses is a source of water pollution which should not be overlooked.

3. In most instances, fisheries have been the victim of water pollution. But in recent years, as the number of fish farms increase, pollution of water areas surrounding fish farms caused by leftover feed and excrete of fish is becoming serious.

Water Pollution by Urban Activity

In a sense, the existence of human beings itself is a source of water pollution. In primitive ages, the purifying capacity of nature was much greater than the rate of water pollution caused by human living. But as time passed, men began to settle near water areas for the convenience of living; then they developed various industries which provided them with the foundation of civilized life. The emergence of various industries coupled with the development of civilization has constantly increased the quantity of waste water containing pollutants.

On the other hand, the increasing concentration of the population has also invited the concentration of pollution sources. As a result, the rate of environmental pollution has exceeded the rate of natural purification.

If all the urban activities of human beings related with water pollution are viewed as a metabolic system, it can be outlined as follows:

Water	Waste water
Energies	Human waste and municipal waste
Substances	Waste gas and waste heat

Municipal wastes such as waste water, human waste, household waste, waste gas and waste heat are the causes of water pollution.

The type of water pollution in a city varies largely depending on the characteristics of the city, lifestyle of the inhabitants, degree of civilization, degree of development of sewage treatment plants or sewerage systems. Although, it is very difficult to investigate these conditions in detail, they can be estimated by investigating the following:

(i) Population of the city: This is the most essential datum to be obtained first.

(ii) Consumption of water and quantity of waste water: where a city water system is well established, it is relatively easy to obtain data concerning the consumption of water. For example, the consumption of city water in a modern civilized city is said to be about 500 litre/man/day. However, it is usually difficult to know the quantity of waste water even if it is only an estimate.

(iii) Condition of sewage treatment: This can be measured in terms of the availability of sewage treatment plants and sewage works including their capacities and efficiency.

(iv) Condition of municipal or public facilities and private facilities. In a large city, there are a number of public facilities and various other facilities, and all these facilities can be the sources of water pollution, such as: (public facilities) Government and municipal offices, schools, hospitals, public bath houses, public lavatories, markets, ports, power plants, city gas facilities, sewage treatment plants, etc.

(Non-public facilities) Factories, stores, offices, gas stations, restaurants, bars, hotels, etc.

(v) Condition of sewage: Municipal sewage mainly consists of a mixture of human waste, household sewage and wastes from other facilities, and its condition varies depending on the characteristics of the city, degree of civilization, precipitation and the condition of road surface.

The quantity and condition of waste water resulting from human activity, mainly consisting of household waste water, generally vary depending on the time in a day; however, the "product of the concentration multiplied by the quantity" tends to remain within a relatively fixed range. Such a value is defined as "pollution loading amount", and is used as the basic data for the sewage treatment planning.

Table 8.0: Standard values of BOD, SS and pollutant load of sewage and human waste

Type of waste water	*Load, equivalent, concentration, etc., of BOD and SS*
Human waste from flush toilets (including those from septic tanks for sewage)	(1) BOD equivalent per person per day: 13 g (BOD: 260 mg/1)
	(2) SS equivalent per person per day: 26 g (SS: 520 mg/1)
	(3) Flushing, water per person per day: 50 g
Household waste water (not including rainwater and special waste water)	(1) BOD equivalent per person per day: 40 g (BOD: 200 mg/1)
	(2) SS equivalent per person per day: 50g (SS-950 mg/1)
	(3) Quantity of waste water per person per day: 200 mg/1

Other typical waste waters	(1) Average BOD of waste water from office building; about 80 mg/1
	(2) Average BOD of waste water from hotels about 100 mg/1 Contd.
	(3) Average BOD of waste water from hospitals about 100 mg/1
	(4) Average BOD of waste water from restaurants and catering facilities: about 400 mg/1

The most conspicuous characteristics of municipal sewage are that the organic content (indicates as BOD) is extremely high the contents of ammoniacal and organic nitrogen, phosphates and sulfices are considerably high and normally, hazardous substances are not contained.

Industrial waste water tends to contain hazardous substances; therefore, its discharge into the municipal sewerage system should be strictly controlled. This is necessary because ordinary sewage treatment plants not only do not have an equipment to remove hazardous substances, but also there is a possibility of getting its activated sludge processing function damaged by such hazardous substances.

The modern sewage treatment technique is capable of removing organic pollutants very efficiently, but it is difficult economically to remove nitrogen and phosphorus contents efficiently with existing facilities. For example, only about 30% of the nitrogen and phosphorus contents can be removed through the currently employed two-step treatment method, which mainly applies the activated sludge method, and the remainder of such contents causes the eutrophication of water areas.

Anyhow, all of the above mentioned techniques are possible where the municipal sewerage system and sewage treatment plants are well established, and the quantity and quality of waste water can be measured effectively. However, where sewage treatment plants are not available, or where waste water is discharged directly into the hydrosphere without passing a sewage treatment plant, it is almost impossible to measure the quality and quantity of waste water.

In case waste water is directly discharged into the hydrosphere, and if the pathogenic bacteria of an infectious disease is contained in the human waste, such bacteria would be discharged into the hydrosphere in active condition thereby eventually spreading the infectious disease through the water system. The degree of such a danger varies depending on the quality of sanitation facilities of individual cities.

Water Pollution by Manufacturing Industry

Of water pollutions with which human activities are concerned, the manufacturing industry is the largest in scale, and most diversified.

The characteristics of the water pollution of manufacturing industry can be summarized as follows:

1. Unlike the mining industry and agriculture, the discharging condition of waste water from manufacturing industries is relatively simple because it is rarely affected by complicated conditions of the natural environment. For example, waste water of the manufacturing industry is discharged concentratively from a fixed outlet, which is called "point source pollutions". Besides, waste water from the manufacturing industry is generated as a result of the production activity which is 100% artificial, and thus, the manufacturing industry can be considered as the source of water pollution which is most easy to catch and control in this point.
2. Sources of water pollution from manufacturing industry are relatively easy trace back through scientific method. But their control is not always easy because it involves complex social and political problems. Furthermore, tracing such point sources of pollution is not always easy because the interests of individual manufacturing companies can be affected by pollution control measures.

3. In contrast to the relatively simple form of discharging process, the comprehension and quantity of waste water vary widely depending on the type and scale of individual factories, change from time to time, and its influence also take.

4. The influence of waste water on the environment varies depending on the geographical location of the factory. For example,

   (i) In the case of an industrial conglomerate or large industrial complex, the influence of waste water on the environment is extremely large, but as far as such manufacturing facilities are concentrated in one area based on the results of careful environmental assessment, waste water from these facilities can be controlled in the most ideal form;

   (ii) In case of factories are located in commercial or residential areas of a city, the location of this type is primarily undesirable. Actually, however, there are a number of such cases because of historical reasons, or geographical advantages concerning the purchase of raw materials, transportation of products, or access to the labour market. In Japan, locations within easy access to the source of industrial water are decreasing in these areas. As a result, factories that recycle water are increasing; small scale manufacturing companies establish cooperative associations and are promoting the construction of joint waste water treatment facilities;

   (iii) Location of the manufacturing factory is whether mountainous area, plain or coastal region is also an important factor to be considered in assessing the influence of waste water on the environment. Especially, in the case of a water consuming type of manufacturing industry, its location has to be decided depending on accessibility to an industrial water source. On the other hand, the condition of the river or the sea into which waste water of the factory has to be discharged also has a great influence on the form and type of environmental pollution.

Characteristics of waste water of Main Industries

As discussed in the forgoing, the composition of industrial waste water is complex, and it varies largely according to the difference in the manufacturing process even among the same type of manufacturing factories. Furthermore, the manufacturing industry is constantly subject to the impact to technological innovations which occur at increasingly short intervals. For example, in Japan, the technique for recycling of waste water is making rapid progress. Manufacturing companies were compelled to do so due to the increasingly strict pollution control requirements and the increasing shortage of industrial water. As a result, manufacturing companies are also compelled to alter conventional manufacturing processes. It is rather difficult to enumerate' the characteristics of waste water of each industry. General characteristics of waste water according to the type of manufacturing industry are as follows:

(1) Food Processing Industry

A common characteristic of waste water from all types of food processing industries is the high concentration of organic content. Another common characteristic is the high concentration of solid, semisolid and suspended solid contents in waste water discharged from raw material treatment processes. The quality and quantity of waste water depending largely on the season mainly due to the variation of raw materials to be processed. Materials contained in waste water can be divided roughly into carbohydrate, protein, fat and a mixture of them. In many instances, waste water contains oil, nitrogen and phosphorous. In the case of starch producing, super refining and brewing industries, the BOD level of waste water is especially high.

(2) Textile Industry

There is much difference in the content of waste water between natural and chemical textile industries. Waste water from the wool industry contains high concentrations of BOD, fat and alkali. Waste water from dyeing discharges various pollutants such as dyes, auxiliary additives and other chemicals.

(3) Paper and Pulp Making Industries

These are typical water-consuming industries and the levels of COD and suspended solids in waste water are extremely high. Especially, waste water from the digesting process contains all components of wood except cellulose and has high COD and dense colour, so that it is necessary to collect and treat most of the content to recover. A large quantity of waste paper fibre is also discharged as suspended solids together with the waste water.

(4) Petroleum Refining Industry

The main component of pollutants is spilled oil; but waste water also contains hydrogen sulfide, other sulfides, ammonia, mercaptan, and phenols, etc.

(5) Petrochemical Industry

The waste water contains hydrocarbons, various organic compounds and catalysts, and its content varies largely depending on the type of product.

(6) Iron and Steel Making Industry

This industry is also a water-consuming industry. Waste water from the cooling and cleaning, processes for coke furnace gas contains ammonia, cyanide, phenols, etc. Besides, there is waste water from the dust collecting process of each furnace which contains suspended solids (coke, dust and ore), and that from the pickling process which contains acid, iron and oil.

(7) Electroplating Industry

Many of the raw materials and chemicals used in this industry contain toxic substances, and thus the waste water contains various heavy metals (cadmium, zinc, copper etc.), cyanide, hexavalent chromium, acids and alkalis.

(8) Leather Industry

The levels of BOD is high in waste water which comes from tannin used for tanning and raws. It also contains chromium used for tanning and lime for pretreatment, suspended solids and dyes.

(9) Steam and Nuclear Power Plants

These power plants discharge large quantities of effluent water from the cooling process. In Japan, all the large power plants are located close to the sea, and these plants are using sea water for cooling. The temperature of waste cooling water is normally high, and such waste water can cause thermal pollution or the elevation of temperature to the vicinity of discharging outlet. No other pollution is considered.

Water Pollution by Other Human Activities

Aside from the various human activities already discussed in previous sections, there are other types of human activities which also cause water pollution, and such activities sometimes become social problems as sources of environmental pollution. These types of human activities are as follows:

(1) Construction work

Construction works, especially those large-scale land development projects or forestry conservation and flood control projects aimed at preventing natural disasters, sometimes can cause water pollution. Water pollution caused by such construction projects can be divided into those occurring incidental to the progress of construction works and those occurring due to the change in natural conditions as a result of the construction works. The latter is more important than the former. Construction works for a dam, river improvement (including those for the alteration of basins), port facilities, creation and reclamation of land and gravel gathering from river-bed are enumerated as examples of the above-mentioned causes.

(2) Various tertiary industries

Various tertiary industries can be considered to cause water pollution such as cleaning and laundry businesses, public bath houses, hotels, printing business, photographic film development businesses, hospitals, laboratories or research institutes, tourist businesses, transportation facilities and so forth. An increase in the load of municipal waste water as a result of the increase in the working population of cities also has become one of the factors directly related to water pollution. Of various transportations, ships, especially tankers which discharge waste cleaning water, have come to the fore as sources of water pollution in coastal regions.

(3) Errors, accidents and disasters

Accidental damages to public sewage works or drainage of industrial effluent, as well as poor planning of sewage treatment plants, and erroneous operations of these facilities can also cause water pollution. Speaking of water pollution caused by accidents, pollution of the ocean due to spilling of oil resulting from the collision of tankers on the high seas has become an object of public attention in recent years.

By now we have taken a look at conditions of water pollution caused by human activities classifying by the type of activity. So as we look at water pollution from a different angle, we may classify and explain it from the standpoint of its physical and chemical characteristics.

(1) Suspended solids (Abbreviated as "SS")

The causes what is called the "turbidity of water". In general, suspended solids are defined as those having a diameter of over lu. Waste water from cement manufacturing industry and some of the mines can be enumerated as examples of waste water containing inorganic SS. Municipal sewage and waste water from kraft pulp factories are typical examples of those containing organic SS.

(2) Organic matter

Organic matter is usually measured as BOD or COD. Organic matter consumes the dissolved oxygen in water, and eventually destroys the ecosystem of the hydrosphere. This is the main content of not only municipal waste water, but also industrial waste water such as those from food processing, chemical, pulp and leather industries.

(3) Inorganic matter

Inorganic matter comprises acids, alkalis, various salts, metals, etc. Waste water containing such substances are those from mines, chemical, pulp, electroplating and other industries.

(4) Inorganic and organic toxic substances

Toxic substances are defined as chemical substances even traces of which adversely affect either the human body or the living environment.

(a) *Inorganic toxic substances :* Comprises metallic substances such as cadmium, lead, hexavalent chromium, mercury, etc., and non-metallic ones such as cyanide and arsenic; and contained in various industrial and mining waste water.

(b) *Organic toxic substances:* The modern chemical industry has created a number of new synthetic substances and their special characteristics to utilize for developing new agricultural chemicals and medicines, etc. However, these synthetic substances adversely affect the environment when they are discharged into it. Many of these substances consist of organic compounds containing phosphorus, chlorine, mercury, etc.

(5) Oil and fat (Hexane Extractable Component)

Comprise vegetable oils, animals fats and mineral oils. Oils and fats are found in municipal waste water and in waste water from food, oil and fat processing industries, etc. While mineral oils are found in the waste water from oil drilling, petroleum refining, petrochemical, iron and steel making and other industries. They are also found in waste water from ships. In Japanese standard method the content of oils and fats is measured quantitatively by extracting them with hexane and is called "hexane extractable component".

(6) Substances contributing to Eutrophication (Nitrogen Compounds and Phosphates)

Salts such as nitrogen and phosphoric acid compounds (NH_4, NO_2, and PO_4) exist in small quantities evta in natural clean water, and the substances are ingested by organisms as nutrient salts; but they also contribute to the eutrophication of water areas when their quantities exceed certain levels, these substances are discharged contained in municipal waste water, surplus of fertilizers, some industrial waste water and synthetic detergents.

(7) Sulfur compounds

Hydrogen sulfide (H_2S) is a typical sulfide that causes water pollution, and is formed in municipal waste water, human waste, waste water from the digesting process of pulp and waste water from gas plants. Hydrogen sulfide gives an offensive odor, and corrodes the facilities and buildings located near the passage of waste water because of its strong reducing power. The content of hydrogen sulfide indicates anaerobic index of the water area. In the volcanic zone, hydrogen sulfide frequently occurs naturally. Sulfur oxides are main pollutants in air and fall to the earth as "acid rain".

(8) Coloring and odor-giving substances

Even if not directly harmful to human health, coloring or odor giving substances in water areas can seriously influence the living environment of people.

Such substances are found in the waste water from dyeing, chemical, agricultural product processing, marine product processing and other industries.

(9) Coliform group

Colon bacillus itself does not have any pathogenic nature, but it is used as an important index to measure pollution by human waste.

Water pollution due to human excrete is caused mainly by the lack of a proper municipal sewerage system and is important since there is a possibility to contain pathogenic organisms.

(10) Radioactive substances

Waste water from research institutes or hospitals that use isotopes, nuclear fuel treatment plants and nuclear power plants are considered to include radioactive substances.

(11) Warm and hot waste waters

Warm or hot waste water consisting of use cooling water discharged from factories and steam or nuclear power plants can significantly influence the water area as part of the living environment when it is discharged in large quantities. However, when the temperature of such waste water is controlled properly it can be utilized for fish farming or district heating.

Table 8.1 shows the major sources activities that cause water pollution.

Table 8.1: Major causes and sources of water pollution

Cause	*Source (industries and others)*
BOD Content	Food processing industry (canned goods, meat and dairy product (organic matter) processing, marine product processing, starch processing), brewery, sugar making, wool processing (wool washing), pulp and paper industry, leather industry, municipal sewerage night soil treatment plant. Suspended solids (Organic SS) Food processing, paper and leather industries, coal (SS) washing, municipal sewerage. (Inorganic SS) Ceramic industry, gravel yard, quarrying industry, mining industry. Oils and fats (Oils, Food processing, petrochemical, iron and steel making, metal fats, mineral oils) finishing, wool processing (Wool washing) industries. Acids or Alkalis Mining, paper making, pulp, electroplating, metallurgical, chemical, textile and leather industries. Ammonia Gas and coke plants. Phenol Gas and coke plants, synthetic resin and other organic chemical industries.
Sulphides	Pulp and gas industries, Sewage treatment plants, dyeing and leather industries.

Color	Dyestuff-manufacturing, dyeing, paint manufacturing, leather, chemical industries, night soil treatment plants.
Odor	Chemical, pulp, marine product processing, meat processing, sugar making, leather industries, sewage and night soil treatment plants.
Detergent	(Surface Municipal sewerage, textile industry, laundry active agent).
High-temperature	Steam and nuclear power water plants.

Eutrophication

Phosphorus and nitrogen compounds (phosphates, ammonium salts, nitrites and nitrates, etc.) are called nutrient salts which are essential to the growth of phytoplanktons and algae participating in the primary production. However if they are excessively supplied together with organic substances, the production activities become too vigorous due to over nutrition, ultimately changing the aquatic ecosystem. This phenomenon is called eutrophication. Eutrophication of lakes is not an extraordinary phenomenon but one which occurs and slowly proceeds even in natural not polluted ecosystem. However, eutrophication in polluted sea and lakes progresses very quickly due to rapid increase of organic substances and nutrient salts, unusual growth of specific organisms and thus destroying the normal aquatic ecosystem. This type of eutrophication is often observed in closed water areas such as lakes and deeply recessed bays; the results is colouring and smelling of the drinking water taken from the lakes, lowered clarity of lake water, or the occurrence of the so called "red tide" in sea and lakes as a visible and conspicuous phenomenon. The red tide is caused by unusual propagation of certain kinds of planktons at such a terrific speed as to callings to colour of water.

Consequently, many environmental trouble occuring such as death of fishes due to clogging of gills with planktons, oxygen deficiency over the polluted area, and generation of poisonous substances.

In Japan, this phenomenon has often been observed in the inland Sea and other closed water areas. The true causes have not yet been determined exactly, but it is generally accepted that phosphorus and nitrogen compounds contained in urban and industrial waste waters are playing an important role.

Contamination of bottom sediments

River, lake and sea bottoms are usually covered with a certain depth of muddy sediments. These sediments contain remain; and excrete of aquatic organisms, other than the earth and sand carried over by rivers. Suspended solids in water tend to sediment remarkably in lakes, slow-flowing rivers, estuaries where the river stream is stagnated by tidal currents, and in closed sea areas.

Cases are also reported where some highly poisonous heavy metals and sparingly soluble organic substances, which had settled on the bottom of water and fixed as sediment over a long period, where dissolved again or raised under some conditions, causing serious hazards. A typical example is mercury in the bottom sediment of the Minamata Bay. It was about twenty years ago when mercury caused the Minamata disease, but mercury hazards are still continuing even after the discharge of mercury from industrial waste water completely stopped.

At the bottoms of lakes and closed areas with high organic loads, dissolved oxygen is consumed rapidly during the putrefaction of organic substances contained in the bottom mud. If vertical mixing of water is insufficient due to stratification, oxygen dissolved on the surface of water from the atmospheric air cannot reach the bottom. In addition, poor clarity of water weakens sufficient penetration of the sun beam and significantly suppresses photosynthetic reactions in the bottom water layer. Under these conditions, the dissolved oxygen in the bottom water mass will decrease, ultimately leading to the zero oxygen state. As a result, fishes in that area will be seriously affected and the surrounding ecosystem may be materially destroyed.

Other causes are also known where elusion of phosphorus and nitrogen from bottom sediments in closed sea areas may be responsible for eutrophication and the resultant occurrence of red tide. Water bottoms often provide favourable spawning grounds for fishes; anaerobic conditions of bottom sediments destroy hatching in many causes.

A survey of bottom sediments provides the environmental information of the neighboring water area in the past over a considerably long period. Thus, it is possible to obtain new findings on the progress of pollution not expected from water quality investigations.

Oil Pollution in oceans

Modern civilization relies in a large measure on petroleum in all aspects. A huge amount of crude oil is carried by supertankers on oceans throughout the world, stored in tanks at terminal stations, and processed into various petroleum products at petrochemical complexes. Many terminal stations and petrochemical complexes are located in coastal zones. Probably a considerable amount of oil is always flowing into the ocean including slight leaks from various ground facilities, leakage from tankers during loading and unloading, and discharge of waste oil used on land. Besides, large scale oil spills due to maritime accidents have often occurred.

Oil split out to the sea tends to spread on the sea surface as a thin layer. Volatile components gradually evaporate and some of them dissolve into water (benzene, for example, dissolves in seawater to a concentration of about 820 ppm at ordinary temperature). The oil membrane thus formed on the sea surface disturbs normal exchange of gases between the seawater and the atmospheric air, and also interrupts the penetration of sun beams to retard the growth of planktons.

The oil gradually becomes thick as the volatile components evaporate; the resultant viscous heavy oil is in or on the seawater, or gather shell fishes and sand on the sea bottom to form oil balls, which drift over long distance and coast shore. The split oil is degraded by some bacteria though very slowly. However, if it drifts faster than it decomposes, pollution will proceed. Heavy oil sometimes sticks to the bodies of seafowls and fishes; particularly, larvas are seriously damaged by oil deposition. Although not so critical, causes are often encountered when fishes cannot be used for food because of the smell of heavy oil.

Pic. 8.1 A Diverse Range of Nature : An eerie stillness over Tsomgo lake.

9

NOISE POLLUTION

Noise has been described as a sound without agreeable musical quality, or as an unwanted or undesired sound. In many countries comparatively little attention is paid to noise, despite its importance in the urban and industrial scene. It is not new hazards, and people residing in urban areas and many occupationally exposed workers develop hearing losses which may be severe, though fortunately got invariably so. There is much to suggest that this hazard is increasingly prejudicing social life and in some instances earning capacity. The prevention of 'noise-induced hearing loss, therefore, should be accorded top priority. This calls for correct assessment of the noise problem and application of various administrative, engineering and medical control measures.

Noise is noless a pollutant than the toxic chemicals in the environment. As a result of increasing mechanisation, the use of increasingly voluminous and complicated machinery and equipment and the stepping up of the pace of production, noise is becoming an increasingly widespread and serious source of discomfort and danger.

Noise may be defined in a number of ways. The definition given the ILO convention No. 148 is: "the term" noise; covers all sound which can result in hearing impairment or be harmful to health or otherwise dangerous.

The Mechanism of Hearing

The ear is made up of three parts: the outer, middle and inner ear (Fig. 9.1). The outer ear extends the ear-drum, which is at the bottom of an oval cylindrical canal which directs sound to the drum. It has a slight S-shaped curve which sometimes causes difficulty with the insertion of ear-plugs; its diameter varies at different points, and it is narrower at the inner end. In talking, chewing or yawning the shape of the outer third of the canal changes, so tending to displace even well-fitting plugs. The elliptical eardrum is thin and transparent and rendered taut by its attachment to the malleus bone.

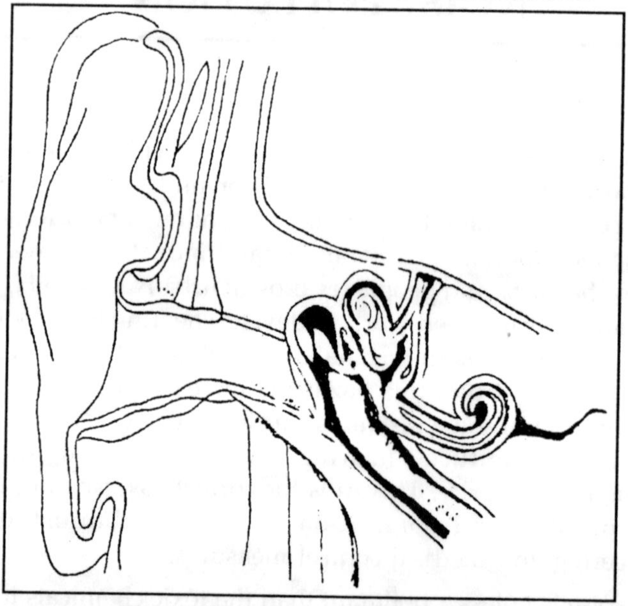

Fig. 9.1 Schematic diagram of Ear

The middle ear contains a chain of three small movable bones called the auditery ossicles-malleus, incus and stapes- and two minute muscles. The Eustachian rube connects the middle-ear cavity with the pharynx, so as to equalise air pressure on both sides of the drum in normal health; this is necessary if the latter is to vibrate when sound impings on its surface. The extent to which the drum moves also depends on other factors; the tone of the nasopharyngel muscles, the presence or absence of middle-ear congestion, and the degree of protection afforded by earplugs, etc.

The inner ear, a complex of cavities contained within dense bone, includes three semicircular canals at right-angles to each other and the cochlea. The canals as part of the body's balancing mechanism, need not concern further.

The cochlea is the site of damage when hearing impairment results from exposure to excessive noise. In shape it resembles a snail's shell, with 21/2 turns around a central hollow pillar, the modulous, which carries the nerve fibres leaving the sensory cells.

Sources of Harmful Noise

The major sources of noise pollution in the developing countries include industry, transport and community activities. These sources can be classified as follows:

(a) continuous wide-bad noise (wide frequency range), e.g., background noise in a mechanical workshop;

(b) continuous narrow-band noise (sound energy mainly concentration on a small number of frequencies), e.g., circular sawa, planing machines, etc.;

(c) impact noise (single pulses of short duration), e.g., forging, hammering;

(d) repetitive impact noise, e.g., rivetting, trimming etc.

(e) intermittent nose, e.g., road traffic, aircraft noise, community, activities, construction, etc.

Measurement of Sound Levels

The instruments used to measure and evaluate noise include sound level meter rectosopm sound level meter, precision integrating sound level meter, doss meter, calibrator, octave-band analyser, graphic level recorder, etc.

Sound Level Meters

The standard sound level meter is the basic measuring instrument used to evaluate potential noise exposure hazards.

It consists of a microphone, an amplifier and an indicating meter. The meter is small battery powered and it registers the sound level at the microphone in terms of decibels.

Sound level meters are considered according to accuracy or type 2, general purpose instruments. In general, the type 2, instruments are useful only for quickly surveying the sound field to determine the approximate sound level at various locations. The type 1 precision instrument is employed to obtain accurate data for noise control purposes.

Sound level meters indicate sound pressure level in the audible range. When an octave filter set added, it becomes an easily operated portable frequency analyser.

When noise fluctuates widely it is very often advantageous to use precision integration source level meter which continuously displays the Leq measurement in non-steady noise environment can then be carried out until the Leq display is stable. This is particularly useful when making factory noise measurements for the purpose of occupational hearing conservation.

Noise Dose Meters

Noise dose meter is a special type of integrating sound level meter designed specifically to meters the noise exposure of workers.

Table 9.0. Class of Hearing Ability Based on Average Value of Hearing Levels at 500, 1000 and 2000 Hz*

	Class Degree of handicap	*Average hearing levelled B*	*Ability to understand ordinary speech*
A.	Not significant	Less than 25	No significant difficulty with faint speech
B.	Slight	25 to less than 25	Difficulty only with faint speech
C.	Mild	40 to less than 55	Frequent difficulty with normal speech
D.	Marked	55 to less than 70	Frequent difficulty with loud speech

E.	Severe	70 to less than 90	Shouted or amplified speech only understood
F.	Extreme	90	Usually even amplified speech not understood.

These hearing levels in dB, apply to the better ear; if the average for the poorer ear is 25 dB or more than for the better ear, add 5 dB to the during their noise working day. The dose meter, in simple terms, consists of a sound level meter, integrator and indicator. It is self-contained and pocket-sized that it can be worn by the workers while at work and thus measures his actual noise exposure however, much the noise changes or wherever may be in the factory. It takes into account noise not only the noise level but also the length of time worker is exposed to it. The instrument indicated the percentage of allowable exposure or an equivalent dB (A) level.

Above a certain level, noise may activate the sympathetic nervous system and therefore give rise to among other things, an increased secretion of among hormones from the adrenal duly (adrenalin and non-adrenalin) and a change in blood circulation.

Excessive noise may cause dizziness or nystagmus. It can also cause increase in respiratory rate, heartburn, somnolence, sleep disturbances and asthenia.

Magnitude of the Risk

Since noise is becoming an increasingly widespread hazard, it is hardly surprising that the number of cases of occupational diseases is increasing year by year.

Employment in registered factories where potential high noise exposure of workers exists, along with employment in either sectors with noise risk as shown in Table 9.1. The noise levels in many of the operations in these industries far exceed the permissible level of 90 dB for an eight-hour exposure. The actual noise levels prevailing in some of these industries is given in Table 9.2 and Table 9.3 indicates the average percentage of noise induced sensori-neural hearing loss in the eight factories to be 25.5 per cent which is very much higher than the sensori-neural loss (5 per cent) in a non-exposed population.

Studies carried out by the Industrial Medicine Division of the Central Labour Institute in forging plants, hydroelectric power plants, textile mills, heavy engineering industries, etc., revealed that noise levels in many sections of the above industries were above the stipulated level and the workers were suffering from various grades of noise induced hearing loss. Analysis of the hearing level data on workers of a forging unit, reveals the length of exposure to noise certainly influence the extent of damage.

Table 9.5 indicates the overall noise levels recorded in two forgoing units. The hearing level data recorded in the study showed that around 60 per cent of the workers in these operations had a hearing levels exceeding 25 dB, and hence could be considered as affected cases (Table 9.6).

A study in a hydroelectric power generating statistics revealed hearing impairment of varying degrees amongst 60.6 per cent of employees out of a total of 162 examined.

International and National Legislation

The International Labour Conference adopted at 54th (1970) Session a resolution calling inter alia the inclusion of occupational deafness in the list of occupational diseases appended to Convention No. 121.

Table 9.1. Employment in Some Important Noise-affected Industries

	(Figures for 1990 in lakhs)
Cotton textile	11.69
Woollen, silk and synthetic fibres	1.72
Jute, hemp and mista textile	2.64
Wood and wood products, furniture and fixture	133
Iron and steel	150
Foundries	2.13

Transport equipment and parts	3.97
Electrical machinery	2.75
Machinery, machine tools and parts (except electrical machinery)	4.72
Grain mill products	1.84
Mining and quarrying	7.71
Construction	10.32

Calibrators

The accuracy of sound level meters may be checked by using an acoustical calibrator. It consists of a small stable sound source that fits over the microphone and generates predetermined sound level. For example, a pistonphone calibrator separates at 250 Hz and produces a sound level of 124 dB.

Octave-bund Analysers

A sound spectrum analysis is conducted to determine how sound energies distributed with frequency. Several types of spectrum analysers are variable. The octave-filter set is compact, portable unit which contains eleven band-pass filters for active analysis. Filter centre frequencies are 31.5 Hz to 31.5 KHz arranged according to the preferred frequencies. Filter sets are used in conjunction with the sound level meters. The octave-band analyser is important because, deafness is better controlled with octave-band levels, rather than with over-all noise intensity.

Graphic Level Recorders

Sound level meters have a display from which noise levels can be read directly. Data usually needs to be documented as well as stored for their reference and comparison.

The level recorder can be used in the field to record and documented as well as a final result. It produces a permanent chart record of the levels an applied electrical signal. For noise measurements, this signal is obtained from the output of a sound level meter or band analyser. The graphic level recorder, when connected to the output of a sound level meter records sound pressure level as a function of time.

Harmful Effects of Noise

Noise mainly effects hearing but it may also affect health in other ways.

Effects on Hearing

Non-auditory Effects

The effect of noise on other bodily functions is much less known than is its effect on hearing. The various bodily reactions to loud sounds are mediated by the vegetatives (autonomous) nervous system, which receives its auditory input from the ascending auditory cortex. These reactions therefore are usually to under conscious control and are probably independent of our interpretation of the sound.

Noise limit levels vary from country to country. The ILO code of practice on the protection of workers against noise and vibration recommends the depending on the degree of protection wanted, the following limit values should be determined:

(a) a warning limit value that sets the noise level under which there is very little risk of hearing impairment to a unprotected ear for and;

(b) a danger limit value that sets the noise level above which hearing impairment and deafness may result from an eight-hour daily exposure of an unprotected ear.

Table 9.2. Noise Levels in Some Industries

Industry/operation	*Sound pressure level (dBA)*
Manufacturing and testing units	94-140
Pharmaceutical	94-128
Founding motor manufacturing plant	104-120
Heavy engineering	94-124
Synthetic fibre manufacturing plant	90-117.5

Electronics	87-122
Hydel power plant	92-106
Land transport workshop	90-124
Cotton Textile mills	92-105
Fertiliser Plant	104-118

Table 9.3. Noise Induced Hearing Loss (NIHL) in Different Industries (Percentage)

Industry	*Normals*	*NIHL*	*Other hearing loss*
Electrical	86.5	7.2	7.2
Value Industry	75.0	10.0	15.0
Foundry	8.0	40.0	52.0
Chemicals	46.1	34.1	19.8
Oil Mill	48.2	32.7	19.1
Refinery	62.5	28.2	9.3
Fertilizer	63.3	12.2	17.7
Textiles	31.3	32.6	36.1
Average percentage for all industries	52.6	25.5	21.9

The code of practice indicates that, in the light of present knowledge, the following values may be recommended: a warning limit value of 85 dB (A) and a danger limit value of 90 dB (B). These values are equivalent continuous sound be related to noise measurements such in accordance with the relevant procedures outlines in the code. It adds that:

(a) no matter for how short a time, a worker should not : without appropriate ear protection, enter an area in which the noise level is 115 dB (A) or more;

(b) if there are single isolated bursts of nose which can be above 120 dB (A) "Impulse" or 120 dB(A) "Fast", personal protective equipment should be worn;

(c) no worker should enter an area where the noise level exceeds 140 dB (A).

In 1976, the Government of India had included noise-induced hearing loss as a notifiable disease under the Factories Act, 1948. Under Section 89 of the Act, any medical practitioner who detects noise induced hearing loss in a worker has to report the case to the Chief Inspector of Factories, along with all other relevant information. Failure to do so is a punishable offence. A Government appointed expert committee which went into the problem of compensation for occupationally-affected cases-recommended that noise-induced hearing loss arising out of occupational exposure should also be included in the compensable diseases under the Workmen's Compensable Act (1923). The addition was in Part-B of Schedule IH of this Act, which is enforced by the State Governments. The Model Rules framed under the Factories Act incorporate permissible levels for noise exposure (Table 9.7; and 9.8) and the control measures necessary to implement these.

Measures to Control Noise Pollution

An effective hearing conservation programme should be undertaken where exposure to industrial noise is capable of producing hearing loss. The object is to ensure that an employee's hearing is not affected during his working life to an extent greater than the usually occurring with age, and to preserve it at a level sufficient for normal speech perception.

The criteria for the hearing conservation programme are:

(i) Conservation programme is needed in situations where the hearing loss, as measured by audiometry, is more than 10 dB at 4,000 Hz.

(ii) It is also indicated when the sound level in the working environment is more than 90 dB and there is difficulty to communicate by speech.

(iii) When it is difficult to hear a loud noise at distance of one foot (0.3 m).

(iv) When workers find it difficult to communicate by speech.

(v) When worker experience head noises or ringing in the ears after work, and/or.

(vi) When workers notice a loss of hearing persisting several hours after exposure.

The programme consists of:

(a) noise exposure analysis;

(b) engineering control of noise;

(c) protection of hearing with ear defenders where necessary; and

(d) measurement of hearing by audiometry.

Table 9.4. Impact of Age and Duration (Year) of Exposure of the Hearing Level of Workers in a Forging Unit

Age group	*0-5*	*6-10*	*Exposure (years)* *11-15* *Mean hearing level in dB*	*16-20*
21-25	20.0	28.5	-	-
26-30	25.0	33.1	58.3	-
31-35	22.5	47.4	57.5	-
36-40	-	47.7	59.7	-
41-45	-	40.0	-	
46-50	-	52.5		

The precision sound level meter and the noise dose meter together provide an effective means of identifying personnel who are at risk and for quantifying noise dose.

The Control Methods

1. Keeping the total noise emission from all noise sources below the dose limit in the work area.

2. Regulating spacing between noise sources and between sources and operators. In free-filed conditions the sound level roughly varies with the squares of the distance.
3. Enclosing the noise sources by a sound reducing i.e. attenuating structure which prevents airborne transmission.
4. Reducing structure-borne transmission by isolation of the source using resilient mountings.
5. Damping of vibrating metal structures or by replacement with materials such as wood.
6. Reducing reflected noise by use of absorbent materials on surfaces such as roof, walls and floors.
7. Placing attenuating screens between the operators and the source.
8. Correcting imbalance and vibration by preventive maintenance.

Table 9.5. Over All Noise Levels Records in Two Forging Units

Location/machine	*Overall level d*	*BA in Noise*
Unit 'A'	Unit 'B'	
Forging hammer	102-107	114
Shot blasting	86-94	86
Compressor room	95	94.95 and 98/104

It is important that the above control procedures should take account of thermal balance and ventilation requirements of operations and machinery.

Ear defenders can be used to reduce noise by a substantial amount depending on the type selected and the efficiency of it. The main types are ear plugs consisting of fine glass, wool, mouldable foam, or moulded plastic: ear cups with a soft seal fitted with a spring or adjustable headbond. Comfort, maintenance and clearing are important aspects of use. Workers exposed to high noise hazards involved and the continuous use of suitable ear protectors should be enforced.

Table 9.6. Percentage of Workers having Varying Degrees of Hearing Impairment

Over 26 dB	*Corresponding hearing impairment*	*Percentage of sample affected in Unit 'A'*	*Unit 'B'*
Upto 10	Upto 15	28	27
11-20	16.5-33	21	23
21-30	31.5-45	7	10
31-40	46.5-60	2	4

Table 9.7. Permissible Exposure in Cases of Continuous Noise

Total time of exposure	*Sound pressure level iii dBA*
8	90
6	92
4	95
3	97
?	100
1 ½	102
1	105
1/2	107
½	110
1/4	155

Notes:

1. No exposure in excess of 119 dBA is to be permitted.
2. For any period of exposure falling in between any figure and the next higher of lower figure as indicated in column 1, the permissible sound pressure level is to be determined by extrapolation on a proportionate basis.

Audiometry is an essential part of any hearing-conservation programme. It facilitates work placement, the detection of changes in auditory fit enables hearing loses at 4000 Hz to be detected long before the victim notices any difficulty in conversion, the identification of noise-susceptible persons, and the evaluation of the effectiveness of ear-protectors and noise-control measures.

Table 9.8. Permissible Exposure Levels of Impulsive or Impact Noise

Sound pressure hi, dB	*Permitted number of impulses or impacts per day*
140	100
135	315
130	1,000
125	3,000
120	10,000

Noise:

1. No exposure in excess of 140 dB peak sound pressure level is permitted.
2. For any peak sound pressure level falling in between any figure and the next higher or lower figure as indicated in column 1, the permitted number of impulses or impacts per day is to be determined by extrapolation on a proportionate basis.

An audiometer is used to measure hearing acuity. Measurements may be made with speech or tone signals. Industrial audiometry is limited almost entirely to air-conduction threshold test using pore tones generated by earphones placed over the ears.

10

DOMESTIC POLLUTION

The main concern in characterizing domestic pollution is health. As people spend most of their time indoors in their houses it is important to characterize the level of air pollution inside the residence and their effects on health.

Domestic pollution can be the result of (i) poor indoor air quality (ii) improper water supply due to poor sanitation or nearby industrial effluents and (iii) inadequate solid waste management system. Of these three, only the domestic air pollution is considered in this article. The sources of domestic pollution, measurement of their strength and the levels inside the house for different pollutants are discussed. Potential health hazards due to these pollutants are pointed out.

The composition and the level of domestic air pollutant is determined by the following factors:

(i) indoor sources of pollution (domestic appliances, incense sticks, smoking, etc.);

(ii) outdoor sources of pollution (industries, transport);

(iii) ventilation conditions in the house; and

(iv) meteorological parameters.

The effect of outdoor sources on the indoor pollutant levels in turn depends on their source strength and distance from the meteorological conditions. Indoor sources also ultimately contribute to ambient air quality and the extent of indoor-outdoor exchange decides the indoor pollutant levels. Most important of these parameters are the nature and the strength of the indoor source. In India, indoor sources range from biomass burning chulhas and hukkas in rural and semi-urban areas to LPG stoves and cigarette smoking in towns and cities.

The ventilation conditions in a house depend on the construction and its location in a micro-meteorological condition. The type of kitchens vary with economic class, and are different in rural and urban areas.

Domestic air quality can vary considerably during the day. The pollutant levels in the kitchen will be significantly higher during the cooking period. The living room can have higher levels depending on the number of smokers and the cigarettes. The total exposure and health impact of these levels will depend on the number of occupants.

Source Distribution

The important indoor sources of pollution (fuels and devices) and the major pollutants emitted by them is given in Table 10.1. Use of biogas and natural gas is still on a very small-scale and hence is not with a oven and four burners are also currently being introduced. The air to fuel ratio and the power range of the burners are the two design parameters that affect the emissions from LPG stoves. Kerosene stoves are of two types: pressure and wick. A large number of wick stove designs are available. Wick stoves emit more CO than pressure stoves (Godsch, 1985).

The design parameters for biomass stoves are combustion volume of primary and secondary air, and the power range of the device. It is far more difficult to characterize biomass stoves due to the wide variation in fuel characteristics and stove designs, than it is to characterize the standard designs based on higher quality fuels.

Other Parameters

Table 10.0 illustrates the effect of burner number and pans on CO emissions. The levels are significantly higher when pans are kept on top of burners. Experiments that measure emission of CO directly from burners will under estimate the emissions that will result in kitchens where pans will invariably be used. A similar effect can be expected for other fuel-stove combinations also.

Measured air exchange rates in rural Indian Kitchens vary from 3-20 air changes per hour (Smith et al., 1983). Though these are significantly higher than those in air tight houses in cold western countries, their effectiveness in removing pollutant from the cooking region is perhaps less than having stoves with an effective hood that is vented outside. In cold regions of the country, the pollutant concentrations reach hazardous levels for prolonged periods of time due to poor ventilation and the absence of a chimney.

Table 10.0. Effect of Burner Number and Pan Covering on Coal Levels in a Kitchen with a Gas Cooking Range

Gas Stove Condition	*Concentration (ppm)*
One burner	11
One burner with pan	13
Two burners	12
Two burner with pans	26
Three burners	18
Three burners with pans	29
Four burners	23
Four burners with pans	41

There is a typical diurnal variation in meteorological conditions resulting in higher concentration of particulates in the evening. There are two periods in the year when the concentrations can be higher than at other times. One is during the winter when inversions are common near ground level and the other during the rainy season. In addition, within a room vertical stratification of smoke is observe with lower concentrations being found at levels close to the ground (Qoshi et al., 1985).

The indoor/outdoor ratio of levels of the air pollutants indicates whether sources are predominantly indoor or outdoor. Table illustrates indoor/outdoor ratios for a few pollutants with different indoor sources. A ratio greater than one generally implies the exist are of an indoor source, and vice versa.

Table 10.1. Indoor/Outdoor (1/0) Relationships of Pollutants

Condition	*Pollutant*	*1/0*	*Remarks*
No indoor Source+	SO_2	0.3-0.5	
Moderate ambient levels of SO_2			Deplection of
			SO_2 by biding
No indoor Source+ Low			
Ambient levels of SO_2	SO_2	0.7-0.9	envelop
Electric Stove	NO_2	0.38	in small
Gas/Kerosene Appliance	NO,	>1	
No indoor source	CO	1	unreativell
Biomass and other stoves	CO	>1	gas
Residences with Smokers	RSP	1.7-4.4	
Residences without Smokers	RSP	1.4-1.7	
Gas utilities	B(a)P	0.8	
(no-smoking)			
Gas utilities (smoking)	B(a)P	2.3	
Electric utilities			
(non-smoking)	B(a)P	1.2	
Wood burning stoves	B(a)P	3.6	

CHARACTERIZATION OF POLLUTANTS

Major Pollutants

The type quality of fuel and the common conditions all effect the magnitude of error pollutants. Most fossil fuels have impurity sulphur, ash, and other toxic metals. Bio-gas have relatively less of these but the combustion conditions are poor as compared to others like kerosene, charcoal and L.P.G. Incomplete combustion results in the creation and release of air is not possible to common on the contribution of tobacco smoke to indoor levels. In Table 10.2, CPRIOTARA CPS are improved one hole metal stoves while CPS is a conventional metal stove. The CO and TSP concentrations observed in SVHs compare well with those measured during the cooking sessions in rural kitchens. Concentrations measured in SVH with mixing fans will be higher than those that would be observed in kitchens. Air quality standards have been established that limit pollutants either in the emissions or in ambient air. Emission Standards: Table 10.3 lists Indian Statistical Institute (ISI) standards for domestic appliances using commercial fuels. A hood is installed on these unvented stoves and its ratio is measured by sampling the fuel gas. The heed method may change the combustion and the emission factors in the kitchens may be quite different. The effect of pans is also not accurate for in setting these standards. In India there are emissions standards available for bio-mass combustion.

It is possible to set standards on emission fact and concentration in fuel gas for different pollutants for chimney stoves. In case of unvents stoves there are measurement difficulties.

Wide range of organic compounds of entirely different character that those present in the fuel itself.

The major combustion-related categories of air pollutants which have been studied extensively are:

Total Suspended Particulates (TSP)

Carbon Monoxide (CO)

Table 10.2 : Emission Factors (g/kg) for Different Pollutants from Various Fuel-Source Combinations

Firewood

Ref. Smith (1984) – Butcher et al. (1984 – Ahuja et al. (1986)

Stove :	*Mudstove*	*Small*	*Large stove*	*Three stove*	*CPRI rock*	*TARA*	*CMs*	*CPS*
Method	3	2	2	2	2	4	4	4
TSP	0.1-0.8	4.2-2.13	9.1-21.3	4.2-10.2	0.3-8.3	0.8-1.8	0.9-1.9	1.5-2.4
CO	1.5	72-92	76-101	39-81	86-360	12-22	13-21	4.9-7.9

Crop Residue — ***Cowdung Cakes***

Ref. : Venkataraman et al. (1986) Smith (1984)Ref. : Venkataraman et al. (1986) Smith (1986)

Stove:	CMS	TARA	CPRI	MUDSTOVE**	CPRI	CMS	MUDS
Methods :	4	4	4	4	4	4	3
TSP	0.6-2.9	2.5-7.4	0.7-1.1	35	1.1-5.4	2.5-3.1	0.7-1.8
CO	19-39	44-81	21-25	110	28-69	25-31	3.82

(Contd.)

Coal			***Tobacco***	
Ref. : Smith (1996)			*Ref. Smith (1984)*	
Stove	*SIGRI*	*Source*	*Passive Cigarette*	*Active Cigarette*
Methods:	*3m*	*Method*	–	–
TSP	1.2	TsP	24	14
CO	120	CO	43	17
		B(a)P (mg/kg)	0.06	0.018
		NOx	2.7	350
		HCHO	1.5	0.03

Kerosene	
Ref. TERI (1986)	
Stove:	*Kerosene*
Method:	4
TSP	3.8
CO	42

Hydrocarbons (HC)

Nitrogen Oxides (NOx)

Sulphur Oxides (SOx)

The gaseous pollutant from different sources are same but the TSP compositions depend on the fuel and their combustion conditions. The particulates from coal combustion have a larger fraction of material than those from biomass inorganic combustion.

It would be useful at this stage to explain different measures caused to characterize a source, indoor air quality exposure and doses which ultimately contribute to health effects.

Emissions factor (g/kg)-A fuel-stove combination is characterised by gins of a pollutant per kg of fuel burnt. The source strength is also called an emission factor and measured in kg of a pollutant emitted per hour.

Concentration (mg/m³ or ppm)

1. Concentrations characterize indoor air quality and depend on the source, its strength, its location on the period of operation, and ventilation conditions. Different averaging times can be used to measure concentration, e.g.
2. Background levels will typically have long averaging time whereas concentrations during operation of indoor sources have short duration.
3. Exposures (mg-h/m^3 or ppm-h)—This parameter indicates the extent of exposure of individuals to pollutants. It depends on the time spent by an individual in an environment with that concentration. Exposures are measured either by person monitors or estimated by measuring concentrations in microenvironments and time spents therein.
4. Dose gms/unit of body weight or gm/unit wt of a body organ. This is the amount of pollutant that is actually deposited in the body. Dose depends on

pollutant size and form, routes of exposure age and many other parameters related to our individual metabolism.

5. Exposure Unit (Ug-person-year/m^3)— One EU is equivalent to one person breathing one microgram of pollutant per cubic metre of air for one year. This unit is used to express population exposure, i.e. population times the average exposure.

Emissions and Concentrations

Different measurements methods are used to measure concentrations and calculate emission factors for different pollutants. For vented stoves the EPA-5 method is used but for unvented stoves different methods are used to sample pollutants.

Particulates from coal combustion are mostly inorganic mineral matter (ash), whereas from biomass smoke they are mainly element of carbon (soot) and organic materials (hydrocarbons). The organic frarrion in TSP is found to increase with slower burning rates (Rau & Huntzicker, 1984). Over 40 compounds have been identified in biomass smoke, about half of which are known carcinogens.

The particle size of wood smoke from fireplaces (Dasch, 1992) as well as of wood and dung smoke from village stoves is found to be mostly less than 3 microns in diameter.

Table 10.3. Specification for Appliances

Standard	*No Stove*	*Environment Parameter* CO/CO_2 inexhaust gas
15:4246-1778	Domestic gas stoves for use with LPG	<0.02
15:2980-1979	Non-pressure stove	<0.02
15:10109-1981	Oil pressure stoves (offset burner) Oil pressure stove	<0.02
15:8749-1978	Gobar-Gas stove	CO<500 ppm

Source: ISI Publication for above standards.

Hence, the particles are mostly respirable, and therefore of more concern.

Based on the ambient air quality data it is possible to fix the number of pollution sources in an area so that their contribution to the air basin does not exceed the limits (Kowalezyk and Tombleson, 1985), as has been done in Oregon for woodburning space heating stoves. This may be a difficult proposition to follow in Indian conditions because of limited options for choice of cooking fuels.

Table 10.2 and 10.3 summarize the emission factor and concentration data on different pollutants. Whereas Table 10.5 deals with ambient air quality.

Health Effects

Direct Evidence

Although concern for human health has been the primary motivation in establishing standards for controlling the concentrations of various substances in air and water, demonstration of adverse health effects directly attributable to indoor air quality remains the weakest link in our understanding of domestic pollution. The reasons are not far to seek. Most pollutants have more than one source, each pollutant can cause or aggravate a number of diseases conditions, each of which can be caused by a number of different pollutants. Neither the endpoint nor the sources are unique in most cases. Most experimental studies of health effects at low levels amongst volunteer subjects control all but the pollutant of interest in practice we are exposed to an unknown mixture of pollutants. Except for some studies of complaints associated with the "sick building syndrome" in office settings, there have been very few studies that have attempted to measure dose and effects in the same group of people. When doses are difficult to measure exposures are used as surrogate variables. When even exposures are difficult to measure, concentrations are used as surrogates. An attempt is made to relate reported concentrations, concentrations of concentration and safe concentrations to the adverse healths associated with various pollutants. With the exception of CO_2

Table 10.4 : Indoor Concentrations (mg/m³) of Different Pollutants

Firewood														
Ref.:			*Ahuja et al. (1986)*			*Smith (1986)*	*Smith (1983)*							
Stove:	*TARA*	*TARA*	*CPRI*		*CPS*	*CMS*	*CMS*		*tradi-*	*open stove*	*trad.*	*trad.*		*trad.*
Location:	*SVH*	*SVH (mixing)*	*SVH*		*SVH (mixing)*	*SVH*	*SVH (mixing)*		*tional Kitchen (Ahmedabad)*	*SVH*	*impro-ved Simula-ted cook-ing in Nepal*	*impro-ved Actual cook-ing in Nepal*		*cook-ing Gujarat 1.1-22.5*
TSP	1.6-2.1	1.3-5.2	1.4-11.1	2.5-3.5	1.9-6.7	3.9-5.5	15.8		4.4	—	—	—	—	1.1-22.5
CO	26-44	36-57	34-56	58-84	23-38	32-49	—	—	690	460	340	77	—	
B(a)P						963-1680mg/m³			—	—	—	—	—	62-1360 mg/m³

Firewood		*Crop Residue*				
Ref.:	*Durgaprasad (1986)*		*Smith (1984)*	*Venkatraman et. al. (1986)*		
Stove:	*Improved*	*Traditional*	*mud*	*CMS*	*Tara*	*CPRI*
Location:	*Kitchen*	*Kitchen*	*SVH*	*SVH*	*SVH*	*SVH*
TSP	3.9+3.0	3.6+3.0	1.0-7.2	(mixing)	(mixing)	(mixing)
CO			9.66	2.3-18	7.6-16.3	7.5-13.8
				49.141	109-174	82-133

Kerosene				*LPG*	*Coal*		
Ref.:	*WHO (1984)*	*TERI (1986)*		*WHO (1984)*	*Smith (1986)*		*Dave (1984)*
Stove:	*Cook-stove*	*Nutan*	*Kerosette*	*Stove*	*Traditional*	*Traditional*	*Sigri*
Location:	*Bombay indoor*	*SVH*	*SVH*	*Bombay Flat*	*Kitchen***	*Kitchen*	*Kitchen*
TSP	.084-.196	1.3-2.7	1.1-2.9	0.05-0.1	5.5	24.9	2.11
CO		20.25	22.25				
B(a)P(ng/m³)	1500-2150					1500-2150	

* Uses dung and wood ** based on wood charcoal

levels, higher concentration as all other pollutants listed in Table 10.5 have been reported from indoor settings than are considered safe.

Table 10.5. Ambient Air Quality Standards for India (mg/m^3) 8 Hr Average

Pollutant	*Area A*	*Area B*	*Area C*
TSP	0.50	0.20	0.10
SO_2	0.12	0.08	0.01
CO	5.00	2.00	1.00
NOx	0.12	0.08	0.03

A: industrial and mixed use area
B: all others
C: tourist resorts and national parks, etc.

In what follows we present a sample of studies wherein the researchers have associated adverse health effects with smoke from biomass combustion. In some cases, no indoor provide concentrations were measured, some provide conflicting evidence, and others need to be corroborated.

Over 35% of the rural households surveyed by NCAER (1981) complained of smoky kitchens an eye problems related to smoke. A little over 119 complained of breathing problems as well.

Anecdotal evidence of increased incidence cataract and other eye problems, especially among women comes from Natwara, a village in Madhya Pradesh (Chandra, 1964). The pollutant most responsible for effects on the eye is likely to be formaldehyde whose acute effects win manifest concentrations less than 1.25 mg/m^3.

A study conducted by the National Institute of Occupational Health among women in Ahmedabad found a statistically significant association between cooking with smoky fuels at the incidence of cough, dysopenea and long abnormalities (NIOH, 1980). The participate concentrations they measured ranged for 4.7 to 11.5 mg/m^3.

11

ATMOSPHERE

The atmosphere refers to a protective blanket which is able to nurture life on the earth and protect it from the hostile environment in outer space. It constitutes the source of carbon dioxide for plant photosynthesis and of oxygen for respiration. It provides the nitrogen which nitrogen-fixing bacteria and man's ammonia manufacturing plants employ to produce chemically bound nitrogen essential for life. As a basic component of the hydrologic cycle, the atmosphere transport water from the oceans to land, thereby acting as the condenser in a vast solar-powered still. Unfortunately, the atmosphere also has been a dumping ground for many pollutant materials ranging from sulfur dioxide to aerosol can Freon-a practice which causes damage to vegetation and materials, shortens human life, and possibly alters the characteristics of the atmosphere itself.

The atmosphere forms an insulating blanket around the earth. Without it the temperature at the equator would rise to 180°F during the day and crop as low as 220°F at night. It burns up meteors that would bombard the surface of the earth from space. Without the atmosphere there would be no sound and no flight. There would be no conventional long-distance radio communication, for this is dependent on the electrons in the upper atmosphere. Without air there would be no lightning, no clouds, no wind, no rain, no snow, and no fire. The surface of the earth would be as bleak and sterile as the moon.

The atmosphere serves a vital protective function. It absorbs most of the cosmic rays from outer space and protects living things from their effects. It absorbs most of the electromagnetic radiation from the sun. Only radiation in the wavelength regions of 300 to 2500 nm and 0.01 to 40m get transmitted to any appreciable extent by the atmosphere. The first region consists of near-ultraviolet, visible, and near-infrared radiation; the second consists of radio waves. One segment is the *optical window*, consisting essentially of the visible spectrum of light, from near-ultraviolet to near-infrared. The other is the radio window, consisting of radio waves from about 1 centimetre to 40 metres in length. It is particularly fortunate for life on earth that the atmosphere filters out tissue-damaging ultraviolet radiation below about 300 nm.

The atmosphere has been essential in maintaining the heat balance of the earth. The atmosphere absorbs infrared radiation emitted by the sun. It also absorbs energy re-emitted from earth in the form of infrared radiation. Therefore, it serves as an important heat stabilizing function and prevents the tremendous temperature extremes which occur on other planets and moons lacking substantial atmospheres.

It is impossible to define the limits of the atmosphere because the atmosphere becomes progressively tenuous with increasing distance from the earth. There is no boundary between the atmosphere and the void of outer space. However, 75 per cent of the earth's atmosphere lies within JO miles of the surface and 99 per cent of the atmosphere lies below an altitude of 30 kilometers (19 miles). The total mass has been estimated at 5500 trillion tons.

Composition of the Atmosphere

The components of the atmosphere may be divided somewhat arbitarily into major, and trace constituents. These components expressed for pollution-free dry air at ground level are given in Table 11.0.

Table 11.0. Components in Pollution-Free Dry Air at Ground Level Expressed as per by Volume

Components	*Volume*
Major Components	
Nitrogen	78.08
Oxygen	20.95
Minor Components	
argon	9.34×10^{-1}
carbon dioxide	3.14×10^{-2}
Trace components	
neon	1.818×10^{-3}
helium	5.25×10^{-4}
methane	2×10^{-4}
krypton	1.14×10^{-4}
nitrous oxide	25×10^{-5}
hydrogen	5×10^{-5}
xenon	$0\text{-}8.7 \times 10^{-6}$
sulphur dioxide	$0\text{-}1 \times 10^{-4}$
ozone	$0\text{-}2 \times 10^{-6}$
nitrogen dioxide	$0\text{-}2 \times 10^{-6}$
ammonia	0-trace
carbon monoxide	0-trace
iodine	0-trace

In addition to the components in Table 11.0, air may be having from 0.1 to 5 per cent water on a volume basis, with a normal range of 1 to 3 per cent. The density of the atmosphere decreases sharply with increasing altitude as a consequence of the gas laws and gravity. Over 99 per cent of the mass of the atmosphere is found within approximately 30 km (1 km =- 0.62 miles) of the earth's surface. The total mass of the atmosphere is nearly 5.5 x 1011 tons. Large as it is, this enormous mass has been only approximately one millionth of the earth's total.

The characteristics of the atmosphere have been found to vary greatly, particularly with altitude. Other factors introducing variability include season, latitude, time and even solar activity. Temperatures in the atmosphere may vary from as low as 138°C to over 1700°C. Atmospheric pressure drops from 1.00 arm at sea level to 3.0×10^{-7} arm at 100 km above sea level. Due to these variations, the chemistry of the atmosphere varies greatly with altitude. In addition to temperature differences, the mean free path of species in the atmosphere (mean distance travelled before collision with another particle) increase over many orders of magnitude with increasing altitude. A particle having a mean free path of approximately 1×10^{-6} cm at sea level is having a mean free path which is exceeding 2×10^{6} cm at an altitude of 500 km.

Major Regions of the Atmosphere

The science, of planetary atmosphere is termed as *acronornif.* In addition to dealing with the chemical composition and reactions of these atmospheres, it involves their movement and thermal balance. In order to understand atmospheric-chemistry it becomes, essential first to have some understanding of the physical characteristics of the atmosphere.

It is possible to divide the atmosphere into a number of different regions depending upon the system of classification. The most general classification has been that of the lower atmosphere (up to approximately 50 km) and the upper atmosphere. Until the development of high altitude rockets and earth satellites, only the lower atmosphere was accessible for study. However, use of these devices has caused a quantum jump of man's understanding of atmospheric phenomena in recent years.

The upper region of the atmosphere is having species appreciably different from those found in the lower region. As the lower regions are relatively homogenous in composition, one system of classification divides the atmosphere into regions called the homosphere (having little variation in composition) and the heterosphere (having a high variation in composition).

The system most commonly used is to divide the atmosphere upon the basis of temperature changes. These temperature regions and their characteristics have been summarized in Table 11.1 and have been depicted schematically in Fig. 11.1. In addition to the regions listed, a fifth region, the exosphere, is sometimes defined. Some of the species in the exosphere attain sufficient kinetic energy to break loose from the gravitational field of the earth and go into outer space.

The boundaries of the troposphere have been found to be influenced by the number of factors, including temperature and the nature of the underlying terrestrial surface. Even during a one day period the upper limit of the troposphere has been found to vary by a kilometer or more. The troposphere has been characterized by falling temperature with increasing altitude, as the distance from heat radiating earth increases. In the absence of air pollution the composition of the troposphere has been somewhat homogeneous. This homogeneity has been due largely to mixing by the constantly circulating air masses in the troposphere. The water content of the troposphere however, is extremely variable due to cloud formation, precipitation, and evaporation of water from terrestrial water bodies.

Table 11.1. Major Regions of the Atmosphere and their Characteristics

Region	*Troposphere Range, °C*	*Altitude range, km*	*Significant chemical species*
Troposphere	15 to-56	0 to 11	$N_2 O_2 CO_2 H_2O$
Stratosphere	-56 to-2	11 to 50	O_3
Mesophere	-2 to-92	50 to 85	$O_2 + NO^+$
Thermosphere	-92 to-1200	85 to 500	$O_2 +, O^+, NO^+$

*These figures are values ranging from the lowest part of the region to the highest part. They may vary.

Troposphere: The zone nearest the earth, the troposphere, is named from the green *tropein* (to turn, to rotate, to change). One property that changes in the troposphere is the temperature. It drops to –55° C (-65°F) by the time an altitude of about 16-18 km (10-11 miles) is reached. The rate of temperature drop is about 6.4°C per 1000 metres (3.5°F per 1000 feet) a figure called the *normal environmental lapse* rate. The region through which this lapse rate occurs defines the troposphere. At the equator the troposphere is about 18 kilometers (11 miles) thick, but at the poles it is only about 8 kilometers (5 miles) thick. Essentially all the atmospheres, water vapour is in the troposphere which therefore encloses essentially all the storms and precipitation of the earth. The jet stream occurs at an altitude of 10.5-12 kilometers (35,000-40,000 feet; 6.5-7.5 miles).

The very cold layer at the top of the troposphere is known as the tropopause. Its low temperature and resulting condensation of water to ice particles prevents water from reaching where it would photo dissociate through the action of intense high energy ultraviolet light. If this were to happen, the hydrogen produced would escape the earth's atmosphere. Hence the tropopause acts as a barrier preventing the net loss of hydrogen from the earth with accompanying decomposition of water.

The weather patterns that we are familiar with are all contained within the troposphere, and mixing in the region is fast, of the order of weeks within a hemisphere, and 1 to 2 years between hemispheres. Vertical mixing occurs in the troposphere but movement into the stratosphere is inhibited by the temperature inversion.

Stratosphere: This zone starts at the top of the troposphere, at a thin atmospheric "shell" called the tropopause, where the temperature begins to stay fairly constant, increasing slightly, in going to an altitude of 50 kilometers (31 miles). The stratosphere includes much of the ozone layer. The warming trend in the stratosphere is caused by a cycle of chemical changes in the ozone layer which converts all the incoming high energy ultraviolet rays to heat. If these rays reached the earth's surface, life as we know it would be impossible. The stratopause marks the narrow zone at the top of the stratosphere where the temperature begins to fall again with increasing altitude as the mesosphere is entered.

The stratosphere is characterized by a rising temperature as the altitude increases. The temperature reaches a maximum at the upper limit of the stratosphere. The temperature increases of the stratosphere is due to the presence of ozone, which may reach a level of 10 ppm. by volume in the mid-region of the stratosphere. Ozone absorbs energy in the form of ultraviolet light and causes an increase in temperature.

The stratosphere has been of interest to aeronautical scientists because it is traversed by airplanes; to communication scientists because of radio and television communications; and to air pollution scientists because global transport of pollution, particularly the above ground, atomic bomb tests and of volcanic eruptions, occurs in the stratosphere; and because absorption and scattering of solar energy also occurs in the stratosphere.

The principal chemical species in the stratosphere, N_2 O_2 O_3 and some H_2O, are chemically active because of their interaction with ultraviolet solar radiation. The interaction of a species with radiation, producing chemical changes, is photochemistry. Radiation is absorbed forming an activated species and, provided the conditions are suitable, a new species is formed.

Though most of the energetic radiation from the sun has been filtered out at higher altitudes, some important chemical reactions take place in the troposphere, particularly at the atmosphere/biosphere and atmosphere/lithosphere interfaces. Two of these reactions are photosynthesis, the overall process given by the two equations

$2H_2O \rightarrow O_2 + 4H + 4e^-$ ..(1)

$CO_2 + 4H+ + 4e^- + [CH_2O] + H_2O$ ..(2)

and nitrogen fixation:

$$N_2 \xrightarrow[\text{hydrogenase}]{\text{nitrogenase}} \underset{\text{in the soil}}{NH_2} \rightarrow NO^-_2 \rightarrow NO^-_3 \quad ..(3)$$

Chemical reactions, for example, hydrolysis and oxidation, also occur at tie surface of the earth. The corrosion:

$Fe - 2e \rightarrow Fe^{2+}$ ..(4)

$O_2 + 2H_2O + 4e \rightarrow 4OH^-$ ..(5)

i.e., $2Fe + O_2 + 2H_2O \rightarrow 2Fe^2 + 4OH \rightarrow Fe_2O_3\ X\ H_2O$ air rust ..(6)

Electrical discharges in the atmosphere initiate reaction between nitrogen and oxygen as follows:

$$N_2 + O_2 \xrightarrow{hv} 2NO \xrightarrow{O_2} 2NO_2 \quad ..(7)$$

The far ultraviolet radiation (wavelength <190 mm) is removed before it reaches the stratosphere by atomic nitrogen and oxygen. A considerable portion of longer wavelength ultraviolet, (i.e., below 340 run) is removed in the stratosphere, for example, radiation of wavelength <242 nm promotes the reaction:

$O_2 + hv \rightarrow O + O^*$ ..(8)

The atomic oxygen produced reacts with O_2 to form ozone

$O_1 + O_2 + a \rightarrow O_3 + M^*$ ..(9)

and the ozone absorbs ultraviolet radiation of wavelength <34C nm;

$O, + hv \rightarrow O_3 + O$ ..(10)

followed by the reaction:

$O_3 + O \rightarrow 2O_2$ ..(11)

Hence a cycle of ozone formation and destruction (Chapman mechanism) exists i.e. $3O_{3'} \rightarrow 2O_3$ ..(12)

and $2O_3 \rightarrow 3O_3$ ..(13)

Other reactions can also occur, the principal among which is the reaction of ozone with nitric oxide.

$NO + O_3 \rightarrow NO_2 + O_2$ ..(14)

$NO_2 + O \rightarrow NO + O_2$ ..(15)

$NO_2 + h\upsilon \rightarrow NO + O\lambda$ <400 nm) ..(16)

The overall process may be put as before:

$2O_3 \rightarrow 3O_2$ [twice reaction (14) plus reactions (15 and (16)] ..(17)

and account for removal of approximately 70% ozone. Species HO_x are involved in similar reactions.

The absorption of ultraviolet radiation in the stratosphere prevents virtually an rauiauon below 290 nm, and most between 290 and 320 nm from reaching the earth. Present day biological life has been protected from this radius, a situation that may not continue as gaseous pollutants are added to the atmosphere.

In reaction (9) an excited species, M*, is produced. The energy may be transferred as heat, which would explain rise in temperature within the stratosphere. Infrared radiation also absorbed and re-emitted by ozone adds to the temperature rise.

Mesosphere: The rhesosphere, on top of the stratosphere, extends roughly to 80 kilometres (50 miles), and its temperature drops to between –80 and 90°C at the mesopause, a thin zone where the temperature stabilizes and sun starts to rise again as the thermosphere is ascended.

The principal chemical species in the mesophere are N_2, O_2, O_2^+ and NO^+, and in the thermosphere are N_2, O_2, O_2^+, O and NO^+. The thermosphere is also called the ionosphere as ionization of molecular and atomic species takes place in the lower layers. The layers are responsible for the reflection of radio waves.

Thermosphere: This zone is above 80 kilometres (50 miles) altitude. The temperature becomes very high, but the air has such a low density that it can hold very little heat. Any denser object in the thermosphere will be extremely hot in sunlight but very cold at night.

The temperature falls with increasing altitude in the mesosphere due to a decrease in the levels of the radiation-absorbing species, particularly ozone. In the higher altitudes of the mesosphere and above, molecules and atoms can completely escape the earth's atmosphere. (This region is called the exosphere). The temperature rises to a maximum of approximately 1200°C in the thermosphere due to the absorption of highly energetic radiation of wavelengths less than approximately 200 run by species in this region.

Atmospheric pressure tends to decrease as an approximately exponential function of altitude. Ideally, in the absence of mixing and at constant absolute temperature, T, the pressure at any given height, Ph, is given by the exponential equation

$$P_h = P_O e^{-Mgh/RT} \quad (1)$$

Where P_0 refers to the pressure at zero altitude (sea level), M the average gram molecular weight of air (28.97 grams/mole), *g* the acceleration of gravity (981 cm sec^{-2} at sea level), *h* the altitude in cm, and R the gas constant (8.314×10^7 erg deg^{-7} $mole^{-1}$). For consistency the units, in Equation (1) are given in the c.g.s. system, although height is normally measured in metres or kilometres and it becomes necessary in using Equation (1) to convert height to cm by multiplying by the appropriate factors.

The pressure drops by the factor e-1 for each increase in altitude equal to the *scale height,* defined as RT/Mg. At an average sea level temperature of 28S°K the scale height is 8 X 10^5 cm or 8 km. At an altitude of 8 km the pressure is only about 39% of sea level pressure.

In order to get a better of the variation of pressure with altitude. Equation (1) may be converted to the logarithmic (base 10) form and h expressed in kilometres

$$\log P_h = \log P - \frac{Mgh \times 10^s}{2.3.3RT} \quad ..(2)$$

If the pressure is taken as 1.00 atm at sea level, the equation becomes simply

$$\log P_h = \frac{Mgh \times 10^5}{2.3.3RT} \quad ..(3)$$

In terms of bulk, by far the greatest portion of the atmosphere is confined to the region very close to the surface of the earth. Less than 1 % of atmospheric gases are more than 40 km from the earth's surface. Such an altitude is of course miniscule compared to the radius of the earth.

The zones just discussed may be defined by changes of temperature with altitude. Changes in chemical composition or activity also occurs with altitude. Up to about 80 kilometers (50 miles) the proportions of components given remain nearly constant, and the long zone up to that altitude is sometimes called the *chemosphere*. Within ic is a region called the chemosphere extending from the upper troposphere where a huge number of chemical changes occur powered by solar radiation. Ozone production is one such activity. Above the homosphere lies the heterosphere. For its first 40 kilometer (25 miles), the ionosphere, the very thin concentration of atoms and molecules, exists largely as electrically charged particles called ions. This globe-enveloping band of ions reflects outgoing radio waves back to earth, making radio transmission possible well beyond the horizon. Television waves are not reflected, and a TV transmiter's range does not extend beyond the horizon. Meteorities entering the earth's atmosphere are generally out in the Ionosphere.

Particles in the Atmosphere

Particles constitute common and significant components of the atmosphere, especially the troposphere. Colloidal-sized particles in the atmosphere are called aerosols. Most aerosols from natural sources are having a diameter of less than 0.2 microns. These particles are termed as Aitken particles and are noted for their ability to induce formation of water droplets in supersaturated atmospheres. Aitken particles originate in nature from sea sprays, smokes, dusts, and the evaporation of organic materials from vegetation. Other typical particles of natural origin in the atmosphere include bacteria, fog, pollen-grains, and volcanic ash.

Among the atmospheric phenomena which involve aerosol particles there have been electrification phenomena in the atmosphere, cloud formation and fog formation. Particles help determine the heat balance of the earth's atmosphere by reflecting light. Probably the most important function of particles in the atmosphere has been their action as nuclei for the formation of ice crystals and water droplets. Current efforts at rain making are centred around the addition of condensing particles to atmospheres supersaturated with water vapour. Dry ice was used in early attempts; now silver iodide, which forms huge numbers of very small particles, is employed.

Particles are involved in many chemical reactions in the atmosphere. Neutralization reactions, which take place most readily in solution, may occur in water droplets suspended in the atmosphere. Small particles of metal oxides and carbon are having a catalytic effect on oxidation reactions. Particles may participate in oxidation reactions which are induced by light. It has been shown that zinc oxide surfaces, photosensitized by light, may cause the oxidation of water to hydrogen peroxide.

Particles may vary in number from several hundred per cubic centimeter, in very pure air, to more than 10^s, per cubic centimeter in highly polluted air. Their size lies in the range of 0.1-10m. The particulate mass level has been found to vary from 10 gm/m^3 in clean air to 60-2000 gm/m^3 in the polluted air in urban areas.

In USA nearly 13.4 MT (metric tonnes) of primary particles have been given out of which about 5 MT have been smaller than 3m.

Metal oxides constitute a major group of inorganic particles in the atmosphere. They are originated from the combustion of fuels. Thus, particulate iron oxide is originated in the combustion of Pyrite containing coal:

$$3Fe_2 + 8O_2 \rightarrow Fe_3O_4 + 6SO_2 \qquad (3)$$

A part of the $CaCo_3$ in the ash fraction of coal charged to CaO and gets discharged through the stack

$$CaCO_3 + heat \rightarrow CaO + CO_2 \qquad ..(4)$$

Automobiles have been the main sources of lead particles in the atmosphere. Combustion of leaded gasoline releases lead halides through the exhaust system

$$Pb(C_2 H_5)_4 + O_2 + (halogenated\ scavengers) \rightarrow PbCl_2 + PbBr_2 + PbBrCI + CO_2 + H_2O \qquad ..(5)$$

Aerosol mists are formed from sulphuric acid which are obtained by oxidation of sulphur dioxide which collects water vapour thereby forming small liquid droplets:

$$2SO_2 + O_3 + 2H_2SO_4 \qquad ..(6)$$

Chemical Processes for Formulation of Organic Particulate Matter

Organic particulate matter are formed from a wide variety of sources such as emissions from vegetation and automobiles, combustion of fuel, etc. Typical organic particulate matter, which is obtained in the form of a benzene-extractable fraction from 200 air samples in USA reveals an average formula, $C_{324} H_{48} O_{3.8} SO_{0.083}$ halogen$_{0.065}$ alkoxy$_{012}$. Such particles have been in the 1m size range and have been regarded as potential health hazards.

Polycyclic (polyñuclear) aromatic hydrocarbons (PAH) present in organic particulate matter are having cactnogenic effect. Among these benzo (α–) pyrene is a well characterized substance. PAH compounds commonly are found in urban atmospheres at about the 20μg/m$_3$ level. They are obtained from the pyrolysis of higher paraffins present in fuels and plant material.

High molecular weight paraffins are pyrolysed for getting $C_{10}H_{32}$ which is further pyrolysed. Most of the PAH compounds get absorbed on soot particles.

Soot is obtained as a residue on combustion of fuel in power plants and automobiles. It constitutes nearly 50% of the particulate load in urban areas. Chemically, it refers to a highly condensed product of polynuclear aromatic hydrocarbons; having several thousand interconnected crystallites, i.e., graphitic platelets each possessing nearly about 100 condensed aromnotic rings. The hydrogen content of soot has been 1-3% and oxygen consent about 5-10% due to partial surface oxidation. Due to its large surface area, soot is known to act as a carrier for toxic organics, e.g., benzo (a) pyrene, and toxic trace metals, e.g., beryllium, cadmium, chromium, manganese, nickel, vanadium, etc.

In general, atmospheric particles enter the respiratory tract of human bodies. However, small particles (0.1-lu size) reach the lungs and are retained there, bringing about health hazards. The removal of particulate matter from gas streams is regarded to be an important strategy for air pollution control.

Chemical and Photochemical Reactions in the Atmosphere

One of the primary obstacles encountered in studying atmospheric chemistry is true chemistry generally must deal with incredibly low concentrations. The detection and analysis of reaction products is quite difficult. Simulating high altitude conditions in the laboratory can be extremely hard due to interferences, such as those from species given off from container walls under conditions of very low pressure. Many chemical reactions which need a third body to absorb excess energy occur very slowly in the upper atmosphere where there is a sparse concentration of third bodies, but take place readily in a container whose wall effectively absorbs energy. Container walls may serve as catalysts for some important reactions or they may absorb important species and react chemically with the more reactive ones.

The absorption of light by chemical species can cause reactions which do not otherwise take place under the conditions (particularly the temperature) of the medium in the absence of light, thus photochemical reactions, even in the absence of a chemical catalyst, take place at temperatures much lower than those which otherwise would be required. Such photochemical reactions, induced by intense solar radiation, play a very important role in ascertaining the nature and ultimate fate of a chemical species in the atmosphere.

Electronically excited molecules have been one of the three relatively reactive and unstable species which are encountered in the atmosphere and which are strongly involved with atmospheric chemical processes. The other two species are atoms; or molecular fragments with unshared electrons, called free radicals and ionized atoms or molecular fragments.

Electronically excited molecules are produced when stable molecules absorb energetic electromagnetic radiation in the ultraviolet or visible region of the spectrum. A molecule may possess several possible excited states, but generally ultraviolet and visible radiation are energetic enough to excite molecules only to several of the lowest energy levels. The nature of the excited state may be understood considering the disposition of electrons in a molecule. Most molecules have an even number of electrons. The electrons occupy orbitals, with *a* maximum of two electrons having opposite spin occupying the same orbital. The absorption of light may promote one of these electrons to a vacant orbital of higher energy.

These excited states are energized compared to the ground state and are chemically reactive species.

Two conditions must be met before a photochemical reaction will occur. The first of these laws of photochemistry is that light must be absorbed by the reacting species. If the absorbed light is in the visible region of the sun specimen, the absorbing species is coloured. Coloured NO_2 is the prime example of such a species in the atmosphere. The second law of photochemical process involves activation of the molecule by the absorption of one quantum of light. The energy of one quantum is equal to the product $h\nu$, where h is Planck's constant, 6.62 X 10^{-27} erg sec, and v is the frequency of the absorbed light, sec^{-1}.

An intermolecular energy transfer may take place in which an excited species transfers energy to another species. A subsequent reaction by the second molecule is termed as a photosensitized reaction. It is believed that part of airglow arises from luminescence by excited sodium atoms, Na, formed in the atmosphere by an intermolecular energy transfer involving the following reaction:

$$O_2 + Na \rightarrow O_2 + Na^*$$

In some cases only energy transfer within the molecule occurs,

$$XY^* = \acute{X}Y+$$

a process called intramolecular transfer, where the dagger, +, denotes an excited state different from the initial one. Finally, the excited species may undergo spontaneous isomerization, as in the conversion of O-nitrosobenzaldehyde to O-nitrosobenxoic acid, a reaction used in chemical actinometers to measure exposure to electromagnetic radiation.

The absorption of very energetic radiation may result in the detachment of an electron,

$$N_2 + hv = N_2^+ + e$$

a process called photoionization. Photoionzation often is considered a subcategory of photodissociation in which one of the dissociation products is an electron.

Light absorbed in the infrared region is not sufficiently energetic to break chemical bonds. The receptor molecules gain vibrational and rotational energy. The energy absorbed as infrared radiation ultimately is dissipated as heat and raises the temperature of the atmosphere. The absorption of infrared radiation is very important in the earth's acquiring heat from the sun and in the retention of energy radiated from the earth's surface.

Reaction of Atmospheric Nitrogen

The atmosphere contains 78 % nitrogen by volume and constitutes an inexhaustible reservoir of that essential element.

Unlike oxygen, which is almost completely dissociated to the monatomic form in higher regions of the thermosphere, molecular nitrogen is not readily dissociated by ultraviolet radiation. However, at altitudes exceeding approximately 100 km atomic nitrogen is produced by photochemical reacts;

$$N_2 + h\upsilon \rightarrow N + N \quad ..(1)$$

Other reactions which may produce monatomic nitrogen have been the following:

$$N_2^+ + O \rightarrow NO^+ + N \quad ..(2)$$

$$NO^+ + e \rightarrow N + O \quad ..(3)$$

$$O^+ + N_2 \rightarrow NO^+ + N \quad ..(4)$$

In one region of the ionosphere, the so-called E region, NO^+ has been one or more predominations. A plausible sequence of reactions by which NO^+ is formed as the following:

$$N_2 + h\upsilon = N_2^+ + e \quad ..(5)$$

$$N_2 + O = NO^+ + N \quad ..(6)$$

In the lowest region of the ionosphere, the D region which extends from approximately 50 km in altitude to 85 km, NO^+ is formed by ionizing radiation:

$$NO + h\upsilon = NO^+ + e \quad ..(7)$$

In the lower part of the d region the ionic species N_{2+} is formed through the action of galactic cosmic rays:

$$N_2 + h\upsilon = NO^+ + e \quad ..(8)$$

Pollutant oxides of nitrogen, particularly NO_2 are key species involved in air pollution and the formation of photochemical among. For example, NO_2 is readily dissociated photochemically to NO and reactive atomic oxygen:

$$NO_2 + h\upsilon = NO + O \quad ..(9)$$

This reaction is the most important primary photochemical process involved in among formation.

Reactions of Atmospheric Oxygen

The primary features of the interaction of oxygen among the atmosphere, lithosphere, hydrosphere. The oxygen cycle is critically important in atmospheric chemistry, geochemical transformations, and Life processes.

Oxygen in the troposphere plays a vital role in processes on the earth's surface. Atmospheric oxygen takes part in energy producing reactions, like the burning of fossil fuels.

$$CH_4 \text{ (in natural gas)} + 2O_2 = CO_2 + 2H_2O \quad ..(1)$$

Atmospheric oxygen is used by aerobic organisms in the degradation of organic material.

$$CH_2O + O_2 \xrightarrow{\text{organisms}} CO_2 + H_2O \quad ..(2)$$

Some oxidative weathering processes, like

$$4FeO + O_2 \rightarrow 2Fe_2O_3 \quad ..(3)$$

consume oxygen. Oxygen gets returned to the atmosphere through the photosynthetic reactions of plants:

$$CO_2 + H_2O \xrightarrow{\text{plants}} CH_2O + O_2 \quad ..(4)$$

Due to the extremely rarefied atmosphere and the effects of ionizing radiation oxygen in the upper atmosphere exists in some forms which are quite different from those stable at lower levels. In addition to molecular oxygen, O_2 the upper atmosphere is having oxygen atoms, O; excited oxygen molecules, O_2 ; and ozone, O_3.

Atomic oxygen, O, is stable primarily in the thermosphere where the atmosphere has been so rarefied that the three-body collisions necessary for the chemicals reaction atomic oxygen seldom take place. Atomic oxygen is produced by a photochemical reaction:

$$O_2 + h\nu \rightarrow O + O \quad ..(5)$$

Ultraviolet radiation in the wavelength regions 135 nm to 176 nm, and 240 nm to 260 nm has been most effective in causing reaction (5) to occur. Due to photochemical dissociation, molecular oxygen, O_2, is virtually nonexistent at very high altitudes. At altitudes exceeding about 80 km the average molecular weight of air is lower than the 28.97 g/ mole observed at sea level due to the high concentration of atomic oxygen resulting from photochemical dissociation of O_2. The resulting division of the atmosphere between a lower section with a uniform molecular weight and a higher region with a non-uniform molecular weight forms the basis for classifying two atmospheric regions as the homosphere and heterosphere, respectively, less than 10% of the oxygen in the atmosphere, at altitudes exceeding approximately 400 km is present as O_2.

Atomic oxygen is partially responsible for the phenomenon known as airglow. Excited atomic oxygen, O', given out visible light at wavelength of 636 nm, 630 nm and 558 nm. This species is formed by reactions such as the following:

$$O + O + O \rightarrow O_2 + O^+ \qquad \text{... (6)}$$

Airglow is a very faint electromagnetic radiation which is continuously emitted by the earth's atmosphere. Although the visible component of airglow is extremely weak, the phenomenon is quite intense in the infrared of the spectrum.

Oxygen ions, O^+ may be formed by ultraviolet radiation acting upon oxygen atoms:

$$O + h\upsilon \rightarrow O^+ + e \qquad \text{..(7)}$$

O^+ is the predominant positive ion in some regions of the ionosphere, it may undergo the following reactions to form other important positive ions:

$$O^+ + O_2 \rightarrow O_2^+ + O \qquad \text{..(8)}$$

$$O^+ + N_2 \rightarrow NO^+ + N$$

In intermediate regions of the ionosphere, the species O_2^+ is formed by absorption of ultraviolet radiation all wavelengths of 17 nm to 103 nm. The reaction

$$O_2 + h\upsilon = O_2^+ + e \qquad \text{..(10)}$$

may also be energized by soft X-rays. Reaction (8) and the

$$N_2^+ + O_2 = N_2 + O_2 \quad ..(11)$$

reaction also give rise to O_2^+ in this region.

Ozone, O_3 is an extremely significant oxygen containing species occurring in the stratosphere. Ozone absorbs harmful ultraviolet radiation and serves as a radiation shield protecting living beings on the earth from the effects of excessive amounts of such radiation. Ozone is formed by a photochemical reaction.

$$O_2 + hv \rightarrow O + O \quad ..(12)$$

(where the wavelength of the exciting radiation must be less than 242.4 nm) followed by a three body reaction,

$$O + O_2 + M \rightarrow O_2 + M \quad ..(13)$$

in which M is another species, like a molecule of N_2 or O_r which absorbs the excess energy given off by the reaction and enables the ozone molecule to stay together. The region of maximum ozone concentration occurs within the range of 25 to 30 km high in the stratosphere. In this region the ozone concentration may reach 10 ppm.

Ozone absorbs ultraviolet light very strongly in the region of 220 nm to 330 nm. If this light were not removed by ozone, severe damage would cause to exposed forms to life on the earth. Absorption of electromagnetic radiation by ozone is also responsible for the maximum temperature encountered at an altitude of approximately 50 km, the boundary between the stratosphere and the mesophere. The reason that the maximum temperature takes place at a higher altitude than that of maximum ozone concentration is that ozone is such an effective absorber of ultraviolet light that only a relatively small fraction of the ultraviolet light reaches the lower altitude. One of the primary environmental problems resulting from the high altitude flights of supersonic planes has been that the possibility that exhaust products from these aircrafts cause reactions removing ozone from the atmosphere. Freon from aerosol cans might be having the same effect.

The absorption of ultraviolet light may bring about the dissociation of ozone.

$$O_3 + h\nu \rightarrow O_2 + O \qquad ..(14)$$

or ozone may decompose upon reaction with atomic oxygen: $O_3 + O \rightarrow O_2 + O_2$ ..(15)

Atomic oxygen may also be eliminated by the following reaction:

$$O + O + M \rightarrow O_2 + M \qquad ..(16)$$

Atmospheric Carbon Dioxide

Carbon dioxide is composing only 0.0314 per cent by volume of the atmosphere. It is the non pollutant species of most concern in the atmosphere. Carbon dioxide, along with water vapour, is mainly responsible for the absorption of infrared energy from the sun and radiation of this energy back to the earth's surface. Now, scientists fear that changes in the atmospheric carbon dioxide level will bring about marked changes in the earth's climate resulting in global warming.

In terms of strictly chemical reactions in the atmosphere, carbon dioxide is a relatively insignificant species. It is present in relatively low concentrations and is not thought to participate significantly in photochemical reactions. The infrared radiation which it absorbs is not sufficiently energetic to cause chemical reactions.

Water in the Atmosphere

The water vapour content of the atmosphere has been to vary over a wide range, especially in the lower atmosphere. The normal range has been 1 to 3 per cent on a volume basis, although air may be having as little as 0.1 per cent of water and as much as 5 per cent. Water circulates through the atmosphere in the hydrologic cycle.

Water vapour absorbs infrared radiation, even more than carbon dioxide, thereby determining the heat balance of the earth. Clouds formed from water vapour are able to reflect light from the sun and have a temperature lowering effect.

Gaseous water in the atmosphere is not known to participate appreciably in most atmospheric chemical reactions. However, photochemical reaction between water and ultraviolet light

$$H_2O + hv = H + OH \quad ..(1)$$

is regarded to be a major contributor of atomic hydrogen in the atmosphere. Excited hydroxide radicals, OH*, are formed by the reaction of atomic hydrogen with ozone.

$$H + O_3 - + OH^- + O_2 \quad ..(2)$$

The radicals infrared radiation which ranges into the extremely long wavelength regions of the visible spectrum:

$$OH^* = OH + hv \quad ..(3)$$

This emission has been partially responsible for the phenomenon of airglow.

Condensed water vapour in the form of very small droplets has been of considerable significance in atmospheric chemistry. Pollutant sulfur dioxide gets oxidized to sulfur trioxide in the atmosphere.

$$2SO_2 + O_2 = 2SO_3 \quad ..(4)$$

and the sulfur trioxide gets incorporated into droplets of water

$$SO_2 + H_2O = H_2SO_4 \quad ..(5)$$

to form sulfuric acid. In some industrial areas the rainfall may have a remarkably low pH as a consequence. The harmful effects of some air pollutants—for instance, the corrosion of metal; by acid forming gas-required the presence of water which may come from the atmosphere. The presence of water vapour has an important influence upon pollution-induced fog formation under some circumstances. Water vapour interacting with particular matter in the atmosphere may reduce visibility to undesirable levels through the formation of aerosol particles.

Fog

Fog forms in the atmosphere close to the earth's surface when the relative moisture content reaches 100%, that is when the vapour pressure equals or exceeds saturation point. As the physical processes of cloud formation and fog formation are identical, fog can also be described as 'clouds close to the ground.'

Ground fog is formed in the evening and at night when the radiation balance of the earth is negative, that is, when outward radiation from the earth exceeds inward radiation from the sun. By day the atmosphere near the earth's surface is charged with water vapour, which because of the sharp drop in temperature at night condenses. Fog formation is intensified when, 'owing to a flow of cold air in the valleys, cold air accumulates or when the water content of the ground (in marshland, wet meadows, etc.) is very high. As ground fog occurs in "radiation nights" and is always associated with temperature inversion in the atmosphere near ground level, it is often referred to as "radiation fog."

Mainly in the autumn an inflow of cold air spreads across warm water or warm moist earth and gives rise to dank fog. As in such areas the temperature differences between a cold air and warm water may amount to 10°C (50°F) or more the surface water evaporates quickly and the water vapour condenses again immediately in the cold air. The fog-so formed breaks rapidly, owing to the unstable temperature layers (warm below cold) and takes on the appearance of smoke. A similar pattern is observed in the summer over woods and roads when the sun comes out again. The water evaporates owing to the powerful energy input on the woodland and road surfaces, and condenses again immediately in the rising air.

We speak of *mixing fog* (warm front fog) when warm air is suspended above a layer of cold air close to the ground so that, owing to the interchange and turbulence of the water vapour, the warm air can generate into the cold air below it. As a result of the drop in the temperature of the mixture caused by intermingling in relation to the saturated warm air the saturated moisture pressure will be reduced and the water vapour condensed.

Owing to thermal reaction in the atmosphere under low-lying inversions high level fog is formed. Often it extends downwards and so becomes a ground fog. Although other factors may contribute to its formation, high level fog is primarily a reduction fog.

Advection fog forms; when warm air of sub-tropical origin extends over a cooled layer of air in the warm air zone of a depression. In this case, the warm air is cooled down by a turbulent exchange from ground level upward, so that upon reaching the point of condensation, fog is formed. Such fog formation occurs only in the winter months, when the temperature differences between the sub-tropical and the temperature zone are substantial. In the summer the differences are too small to promote any noteworthy cooling off. Fog formation of this order also occurs when warm moist air passes over fold water surfaces. This phenomenon is common to all cold marine current and occurs mainly in the summer.

Temperature of the Air Layer Near the Ground A factor of far reaching importance of the life processes in the biosphere is the temperature. The temperature, as measured by changes in the volume of mercury in a thermometer, is a reflection of the average kinetic energy in a molecule of air, which is derived essentially from the impact of solar radiation on the earth's surface. The temperature of the layer of air nearest the ground is thus determined by the extent of the solar energy radiated. This in turn is influenced by the composition of the atmosphere, the geographical latitude, the nature of the soil or rock surface, the features of the terrain, etc. The inflow of cold or hot air masses from extensive currents can change the temperature of the layer of air near the ground very substantially.

By day part of the suns rays penetrate the atmosphere to reach the earth's surface, where some are converted into thermal radiation (infrared rays) and returned to the atmosphere. This thermal radiation is absorbed by the atmosphere, which thus becomes richer in energy, i.e., warmer. As most of the energy is imparted to the molecules nearest to the source of energy, that is, the earth, the air immediately above the ground is warmed most. In the temperature zone on a fine summer day with no wind, temperatures of over 21°C (70°F) are recorded in the atmosphere close to the ground. The warm air close to the earth's surface is lighter in specific gravity than the cold air above it, and consequently it tends to rise. During the upward flow, the energy rich molecules impart energy to the surrounding atmosphere, until an energy balance is achieved. Owing to the radiation of solar energy by day, a temperature gradient forms in the layer or air close to the ground. The air temperature falls as the height above the earth's surface increases. This process always occurs during radiation, about it is naturally more marked when air movements are slight. Strong air movements change the temperature conditions in the layer of air near the ground considerably.

At night the thermal radiation from the soil or rock surface of the earth is no longer replenished by solar radiation. As the energy source is lacking, radiation during the night weakens progressively. Consequently the cold air descending towards the ground is not warmed up again and remains as a layer of cold air on the earth's surface. According to the nature of the terrain, the cold air will flow under the pull of gravity into depressions such as valleys, ravines and sink holes to form cold air pockets. Temperature differences of 10°C (SOT) and more between cold air pockets and the warmer zones at higher levels are by no means rare.

Clouds reduce the radiation of the thermal energy considerably, by deflecting it from their underside. Sudden temperature increases of 50°C (44°F) and more observed in the vicinity of a cloud bank. Strong winds also obstruct the formation of a temperature gradient at night. The temperature gradient in the layer of air near the ground is consequently subject to constant reversal between day and night. The conditions that lead to the formation of these gradients can be affected by wind, by the character and structure of the terrain, and by cloud formation and other atmospheric influences.

Evolution of the Atmosphere

It is now widely accepted that the earth's atmosphere originally was very different from its present state, and that the changes were brought about by biological activity.

Nearly 3.5 billion years ago when the first primitive life molecule got formed, the atmosphere was chemically reducing, having primarily methane, ammonia, water vapour, and hydrogen. The atmosphere was bombarded by intense, bondbreaking, ultraviolet light. This light, along, with lightening and radiation from radionuclides, provided the energy to bring about chemical reactions which caused the production of relatively complicated molecules, including even amino acids and sugars.

From the rich chemical mixture in the sea, life molecules got evolved. Initially these very primitive life forms derived their energy from fermentation of organic matter formed by chemical and photochemical processes, but eventually they gained the capability to produce organic matter, "CH_2O", by photosynthesis.

$$CO_2 + H_2O + h\upsilon \rightarrow CH_2O + O_2 \quad ..(1)$$

and the stage was set for the massive biochemical transformation which caused the production of all the atmosphere's oxygen.

The life forms might have been blue-green algae whose limy deposits have been occurred in the oldest of Pre-Cambrian rocks. They were assumed to be first authentic members of the plant world (1400 million years ago). Probably, there was some oxygen which was essential to support the primitive life form. This presumably got formed in the upper atmosphere by the breakdown of water vapour by electrical discharge and solar radiation. The first plant and similar species might be the earth's first mass-producers of free O_2. When the Supply of O_2 got increased more complex plants got evolved and accelerated the process which got stepped up for millions of years, accompanied by an enormous increase in O_2 consuming animals to help maintain the O_2 balance. The present composition of the atmosphere has been virtually the same as it was in the Cambrian age (500 million years ago). But it has been continually changing. The mobiles gases of which it has been composed are always being absorbed by the soil, the rocks and living things or lost into outer space. The supply of atmospheric CO_2 is to be replenished every 4 to 8 years, O_2 every 3000 years and N_2^1 every 100 million years. The cycle of use and renewal of the atmospheric gases depicts an intriguing balance involving soil, air, animals and plants. It is to be noted that all living things have been mainly made of the elements of air; O_2^1, H_2O, C and N.

The oxygen initially produced by photosynthesis was probably quite toxic to primitive life forms. However, much of this oxygen got converted to iron oxides,

$$4Fe^{2+} + O^2 + 4H_2O = 2Fe_2O_3 + 8H^+ \quad ..(2)$$

by reaction with soluble iron (II). This caused the formation of enormous deposits of iron oxides, which provide major evidence for the liberation of free oxygen in the primitive earth atmosphere.

Eventually enzyme systems developed which enabled organisms to mediate the reaction of waste product oxygen with oxidizable organic matter in the sea. Later this mode of waste product disposal was used by organisms for producing energy by respiration, which is now the primary mechanism by which most organisms obtain energy.

It was having an additional benefit in that it enabled the formation of an ozone shield. This ozone shield absorbs bond-rupturing ultraviolet light. With the ozone shield protecting tissue for destruction by high energy ultraviolet radiation, the earth became a much more hospitable environment for life, and life forms were enabled to move from the sea to land.

Pic. 11.1. Discover the natural wealth of a city.

Pic. 11.2. The golf is great, scenic and very affordable at the Dunes, the top public access course in Victoria.

12

THE ENVIRONMENT (PROTECTION) BILL, 1986: A BILL

As bill as passed by Lok Sabha on 8th May, 1986 to provide for the protection and improvement of environment and for matters connected therewith.

Whereas decisions were taken at the United Nations Conference on The Human Environment held at Stockholm in June, 1972, in which India participated, to take appropriate steps for the protection and improvement of human environment.

And whereas it is considered necessary further to implement the decisions aforesaid in so far as they relate to the protection and improvement of environment and the prevention of hazards to human beings, other living creatures, plants and property.

Bill enacted by Parliament in the Thirty-seventh Year of the Republic of India as follows:

CHAPTER I

Preliminary

1. (a) This Act may be called the Environment (Protection) Act, 1986.

(b) It extends to the whole of India.

(c) It shall come into force on such date as the Central Government may, by notification in the Official Gazette, appoint and different dates may be appointed for different provisions of this Act and for different areas.

2. In this Act, unless the context otherwise requires,

(a) "environment" includes water, air and land and the interrelationship which exists among and between water, air and land, and human beings, other living creatures, plants, micro-organism and property;

(b) "environmental pollutant" means any solid, liquid or gaseous substance present in such concentration as may be or tend to be injurious to environment;

(c) "environmental pollution" means the presence in the environment of any environmental pollutant;

(d) "handling" in relation to any substance, means the manufacture, processing, treatment, package storage, transportation, use, collection, destruction, conversion, offering for sale, transfer of the like of such substance;

(e) "hazardous substance" means any substance or preparation which, by reasons of its chemical or physico-chemical properties or handling, is liable to cause harm to human beings, other living creatures, plants, microorganisms, property of the environment;

(f) "occupier", in relation to any factory or premises, means a person who has control over the affairs of the factory or the premises and includes, in relation to any substance, the person in possession of the substance;

(g) "prescribed" means prescribed by rules made under this Act.

CHAPTER II

General Powers of the Central Government

3. (a) Subject to the provisions of this Act, the Central Government shall have the power to take all such measures as it deems necessary or expedient for the purpose of protecting and improving the quality of the environment and preventing, controlling, and abating enviromental pollution.

(b) In particular, and without prejudice to the generality of the provisions of Sub-section (2). Such measures may include measures with respect to all or any of the following matters, namely:

(i) co-ordination of actions by the State Governments, officers and other authorities

(a) under this Act, or the rules made thereunder; or

(b) under any other law for the time being in force which is relatable to the objects of this Act;

(ii) planning and execution of a nation-wide programme for the prevention, control and abatement of environmental pollution;

(iii) laying down standards for the quality of environment in its various aspects;

(iv) laying down standards for emission or discharge of environmental pollutants from various sources whatsoever;

Provided that standards for emission or discharge may be laid down under this clause from different sources having regard to the quality or composition of the emission or discharge of environmental pollutants from such sources;

(v) restriction of areas in which any industries, operations or processes or class of industries, operations or processes shall not be carried out or shall be carried out subject to certain safeguards;

(vi) laying down procedures and safeguards for the prevention of accidents which may cause environmental pollution and remedial measures for such accidents;

(vii) laying down procedures and safeguards for the handling of hazardous substances;

(viii) examination of such manufacturing processes, materials and substances as are likely to cause environmental pollution;

(ix) carrying out and sponsoring investigations and research relating to problems of environmental pollution;

(x) inspection of any premises, plant, equipment machinery, manufacturing or other processes, materials or substances and giving, by order, of such directions to such authorities, officers or persons as it may consider necessary to take steps for the prevention, control and abatement of environmental pollution;

(xi) establishment or recognition of environmental laboratories and institutes to carry out the functions entrusted to such environmental laboratories and institutes under this Act;

(xii) collection and dissemination of information in respect of matters relating to environmental pollution;

(xiii) preparation of manuals, codes or guides relating to the prevention, control and abatement of environmental pollution;

(xiv) such other matters as the Central Government deems necessary or expedient for the purpose of securing the effective implementation of the provisions of this Act.

(3) The Central Government may, if it considers it necessary or expedient so to do for the purposes of this Act, by order, published in the Official Gazette constitute an authority by such name or names as may be specified in the order for the purpose of exercising and performing such of the powers and functions (including the power to issue directions under Section 5) of the Central Government under this Act and for taking measures with respect to such of the matters referred in sub-section (2) as may be mentioned in the order and subject to the supervision and control of the Central Government and the provisions of such order, such authority or authorities may exercise the powers or perform the functions or take the measures so mentioned in the order as if such authority or authorities had been empowered by this Act to exercise those power or perform those functions or take such measures.

4. (a) Without prejudice to the provisions of Sub-section (3) of section 3, the Central Government may appoint officers with such designations as it thinks for the purposes of the Act and many entrust to them such of the powers and functions under this Act as it may deem fit.

(b) The officers appointed under Sub-section (1) shall be subject to the general control and direction of the Central Government or, if so directed by that Government, also of the authority or authorities, if any, constituted under sub-section (3) of section 3 or of any other authority or officers.

5. Not withstanding anything contained in any law but subject to the provisions of this Act, the Central Government in the exercise of its power and

performance of its functions under this Act, may issue directions in writing to any persons, officer or any authority and such person, officer or any authority shall be to imply with such directions.

Fig. 12.1. A Managing Wild Life

Explanation-For the avoidance of doubts, it is thereby declared that the power to issue directions under this section includes the power to direct, the closure prohibition or regulation of any industry, operation or the stoppage or regulation of the supply of electricity or water or any other service.

6. (1) The Central Government may, by notification in the Official Gazette, make rules in respect of all or any of the matters referred to in Section,

(2) In particular and without prejudice to the generality of the foregoing power, such rules may provide for all or any of the following matters namely,

Fig. 12.2. An Ecological Niche

(a) the standards of quality of air, water or soil for various areas and purposes;

(b) the maximum allowable limits of concentration of various environmental pollutants (including noise) for different areas;

(c) the procedures and safeguards for the handling of hazardous substances;

(d) in the prohibition and restrictions on travelling of hazardous substances in different lines;

(e) in the prohibition and restrictions on the location of industries and the carrying on of processes and informations in different areas;

(f) the procedures and safeguards for the prevention of accidents which may cause environmental pollution and for providing for medical measures for such accidents.

Chapter III

Prevention, Control and Abatement of Environmental Pollution

7. No person carrying on any industry, operation or process shall discharge or emit or permit to be discharged or emitted any environmental pollutant in excess of such standards as may be prescribed.

8. No person shall handle or cause to be handled any hazardous substance except in accordance with such procedure and after complying with such safeguards as may be prescribed.

9. (1) Where the discharge of any environmental pollutant in excess of the prescribed standards occurs or is apprehended to occur due to any accident or other unforeseen act or event, the person responsible for such discharge and the person in charge of the place at which such discharge occurs or is apprehended to occur shall be bound to prevent or mitigate the environment pollution caused as result of such discharge and shall also forthwith

(a) intimate the fact of such occurrence or apprehension of such occurrence and

(b) be found, if called upon, to render all assistance, to such authorities or agencies as may be prescribed.

(2) On receipt of information with respect to, the fact or apprehension of an occurrence of the nature referred to in Sub-section (1) whether through intimation under that subsection or otherwise, the authorities or agencies referred

to in sub-section (1) shall, as early as practicable cause, such remedial measures to be taken as, are necessary to prevent or mitigate the environmental pollution.

(3) The expenses, if any, incurred by any authority or agency with respect to the remedial measures referred to in sub-section (2), together with interest (at such reasonable rate as the Government may, by order, fix) from the date when demand for the expenses is made until it is paid may be recovered by such authority or agency from the person concerned as arrears of land revenue or of public demand.

10. (1) Subject to the provisions of this section, any person empowered by the Central Government in this behalf shall have a right to enter, at all reasonable times with such assistance as he considers necessary at any place :

(a) for the purpose of performing any of the functions of the Central Government entrusted to him;

(b) for the purpose of determining whether and if so in what manner, any such functions are to be performed or whether any provisions of this Act or the rules made thereunder or any notice, order, direction or authorisation served, made, given or granted under this Act is being or has been complied with;

(c) for the purpose of examining and testing any equipment, industrial plant, record, register, document or any other material object or for conducting a search of any building in which he has reason to believe that an offence under this Act or the rules made thereunder has been or is being or is about to be committed and for seizing any such equipment, industrial plant, record, register, document or other material object if he has reasons to believe that it may furnish evidence of the commission of an offence punishable under this Act or the rules made thereunder or that such seizure is necessary to prevent or mitigate environment pollution.

(2) Every person carrying on any industry, operation or process or handling any hazardous substance shall be bound to render all assistance to the person empowered by the Central Government under sub-section (1) for carrying out the functions under that sub-section and if he fails to do so without any reasonable cause or excuse, he shall be guilty of an offence under this Act.

(3) If any person wilfully delays or obstructs any person empowered by the Central Government under sub-section (1) in the performance of his functions he shall be guilty of an offence under this Act.

(4) The provisions of the Code of Criminal Procedure, 1973 in relation to the State of Jammu and Kashmir or any area in which that Code is not in force. The provisions of any corresponding law in force in that State or area shall so far as may be, apply to any search or seizure under this section as they apply to any search or seizure made under the authority of a warrant issued under Section 94 of the said Code or as the case may be, under the corresponding provision of the said law.

11. (1) The Central Government or any officer empowered by it in this behalf shall have power to take, for the

purpose of analysis, samples of air, water, soil or other substance from any factory, premises or other place in such manner as may prescribed.

(2) The result of any analysis of a sample taken under sub-section (1) shall not be admissible in evidence in any legal proceeding unless the provisions of sub-sections (3) and (4) are complete with.

(3) Subject to the provisions of sub-section (4) the person taking the sample under sub-section (1) shall

(a) serve on the occupier or his agent or person charge of the place a notice men and there in such form as may be prescribed of his intention to have it so analysed;

(b) in the presence of the occupier or his agent person;

(c) cause the sample to be placed in a container or containers which shall be marked and sealed and shall also be signed both by the person taking the sample and the occupier or his agent or other person;

(d) send without delay, the container or the container to the laboratory established or recognised by the Central Government under Section 12.

(4) When a sample is taken for analysis under sub-section (1) and the person taking the sample serves on the occupier or his agent or person, it notices under clause (a) of sub-section (3), then

(a) in a case where the occupier, his agent or person wilfully absents himself, the person taking the sample shall collect the sample for analysis to be placed in a container or containers which should be marked and sealed and shall also be signed by the person taking the sample; and

(b) in a case where the occupier or his agent or person present at the time of taking the sample refuses to sign the marked and sealed container or containers of the sample as required under clause;

(c) of sub-section (3) the marked and sealed container or containers shall be signed by the person taking the samples, and the container or containers shall be sent without delay by the person taking the sample for analysis to the laboratory established or recognised under section 12 and such person shall inform the Government analyst appointed or recognised under section in writing, about the wilful absence of the person or his agent or person or, as the case may be, refusal to sign the container or containers.

12. (1) The Central Government may, by notification in the Official Gazette

(a) establish one or more environmental laboratories;

(2) Recognise one or more laboratories to carry out the functions entrusted to an environmental laboratory under this Act.

(3) The Central Government may by notification in Official Gazette, make rules specifying

(a) the functions of the environmental laboratory;

(b) the procedure for the submission to the said laboratory of samples of air, water, soil or other substance for analysis or tests, the form of the laboratory report thereon and the fees payable for such report;

(c) such other matters as may be necessary or expedient to enable that laboratory to carry out its functions.

13. The Central Government may by notification in the Official Gazette, appoint or recognise such person as it thinks fit and having the prescribed qualifications to be Government Analysts for the purpose of analysis of samples of air, water, soil or other subs-

tance sent for analysis to any environmental laboratory established or recognised under sub-section (1) of section 12.

14. Any document purporting to be a report signed by a Government Analyst may be used as evidence of the facts stated therein in any proceeding under this Act.

15. (1) Whoever fails to comply with or contravenes any of the provisions of this Act or the rules made or orders or directions issued thereunder, shall in respect of each such failure or contravention be punishable with imprisonment a term which may extend to five years or with in which may extend to one lakh rupees, or with both, and in case the failure or contravention continues, with additional fine which may extend to five thousand rupees for every day during which such failure or contravention continues after the conviction for the first such failure or contravention.

(2) If the failure or contravention referred to in sub-section (1) continues beyond a period of one year after the date of conviction, the offender shall be punishable with imprisonment for a term which may extend to seven years.

16. (1) Where any offence under this Act has been committed by a company, every person who, at the time the offence was committed, was directly in charge of and was responsible to the company for the conduct of the business of the company as well as the company, shall be deemed to be guilty of the offence and shall be liable to be proceeded against and punished accordingly;

Provided that nothing contained in this sub-section shall render any such person liable to any punishment provided in this Act, if he proves that the offence was committed without his knowledge or that he exercised all due diligence to prevent the commission of such offence.

(2) Not withstanding anything contained in sub-section (1), where an offence under this Act has been committed by a company and it is proved that the offence has been committed with the consent or convenience of or is attributable to any neglect on the part of, any director, manager, secretary or other officer of the company, such director, manager, secretary or other officer shall also be deemed to be guilty of that offence and shall be liable to be proceeded against and punished accordingly.

Explanation: For the purposes of this section-(a) 'company' means any body corporate, and include a firm or other association of individuals; and

(b) "director", in relation to a firm means a partner in the firm.

17. (1) Where an office under this Act has been committed by any Department of Government, the Head of the Department shall be deemed to be guilty of the offence and shall be liable to be proceeded against and punished accordingly;

Provided that nothing contained in this section shall render such Head of the Department liable to any punishment if he proves that the offence was committed without his knowledge or that he exercised all due diligence to prevent the commission of such offence.

(2) Not withstanding anything contained in sub-section (1), where an offence under this Act has been committed by a Department of Government and it is proved that the offence has been committed with the consent or convenience of or is attributable to

any neglect on the part of any office other than the Head of the Department, such officer shall also be deemed to be guilty of that offence and shall be liable to be proceeded against and punished accordingly.

CHAPTER IV

Miscellaneous

18. No suit prosecution or other legal proceeding shall lie against the Government or any officer or other employee of the Government or any authority constituted under this Act or any member, officer or other employee of such authority in respect of anything which is done or intended to be done in good faith in pursuance of this Act or the rules made or orders or directions issued thereunder.

19. No court shall take cognizance of any offence under this Act except on a complaint made by

(a) the Central Government or any authority or officer authorised in this behalf by that Government; or

(b) any person who has given notice of not less than sixty days in the manner prescribed, of the alleged offences and of his intention to make a complaint to the Central Government or the authority or officer authorised as aforesaid.

20. The Central Government may, in relation to its functions under this Act, from time to time require any person, officer, State government or other authority to furnish to it or any prescribed authority or officer any reports, returns, statistics, accounts and other information and such person officer, State Government or other authority shall be bound to do so.

21. All the members of the authority constituted, if any under section 3 and all officers and other employees of such authority when acting or purporting to act

in pursuance of any provisions intimation of this Act or the rules made or orders or directions issued thereunder shall be deemed to be public servants within the meaning of section 21 of the Indian Penal Code.

22. No civil court shall have jurisdiction to entertain any suit or proceeding in respect of anything done, action taken or order or direction issued by the Central Government or any other authority or officer in pursuance of any power conferred by or in relation to its or his functions under this Act.

23. Without prejudice to the provisions of sub-section (3) of section 3 the Central Government may, by Notification in the Official Gazette, delegate subject to such conditions and limitations as may be specified in the notification such of its powers and functions under this Act (except the power to constitute an authority under sub-section (3) of section 3 and to make rules under section (25) as it may deem necessary or expedient to any officer, State government or other authority.

24. (1) Subject to the provisions of sub-section the provisions of this Act and the rules or order made therein shall have effect notwithstanding anything inconsistent therewith contained in an enactment other than this Act.

(2) Where any act or omission constitutes an offence punishable under this Act and also under any other Act then the offender found guilty of such offence shall be liable to be punished under the other Act and not under this Act.

25. (1) The Central Government may, by notification in the Official Gazette, make rules carrying out the purposes of this Act.

(2) In particular, and without prejudice to the generality of the foregoing power, such rules provide for all or any of the following matters namely:-

(a) the standards in excess of which environment pollutants shall not be discharged or emitted, section 7;

(b) the procedure in accordance with and the safeguards in compliance with which hazards substances shall be handled or causes to be under section 8;

(c) the authorities or agencies to which intimation of the fact of occurrence or apprehension of occurrence of the discharge of any environment pollutant in excess of the prescribed standard be given and to whom all assistance shall be liable to be rendered under sub-section (1) of section 11;

(d) the manner in which samples of air or other substance for the purpose of analysis be taken under sub-section (1) of section 11;

(e) the form in which notice of intention to be sample analysed shall be served under clause sub-section (3) of Section 11;

(f) the functions of the environmental laboratory the procedure for the submission to such laboratories of samples of air, water, soil and substances for analysis or test; the form of laboratory report; the fees payable for such report and other matters to enable such laboratories carry out their functions under Sub-section (1) of section 12;

(g) the qualifications of Government analyst appointed or recognised for the purpose of analysis of samples of air, water, soil or other substances under Section 13;

(h) the manner in which notice of the offence and of the intention to make a complaint to the Central

Government shall be given under clause (b) of section 19;

(i) the authority of officer to whom any reports, returns, statistics, accounts and other information shall be furnished under section 20;

(j) any other matter which is required to be or may be prescribed.

26. Every rule made under this Act shall be laid, as soon as may be, after it is made, before each House of Parliament while it is in session, for a total period of thirty days which may be comprised in one session or in two or more successive sessions, and if, before the expiry of the session immediately following the session or the successive sessions aforesaid, both Houses agree in making any modification in the rule or both Houses agree that the rule should not be made, the rule shall thereafter have effect only in such modified form or be of no effect, as the case may be; so, however, that any such modification or annulment shall be without prejudice to the validity of anything previously done under that rule.

13

AN INTRODUCTION: BIODIVERSITY

"Biological diversity" means the variability among living organisms from all sources including, inter alia, terrestrial, marine and other aquatic ecosystems and the ecological complexes of which they are part, this includes diversity within species, between species and of ecosystems.

Biodiversity refers to the variety and variability among living organisms and the ecosystem complexes in which they occur. It includes diversity of forms right from the molecular unit to the individual organism, and then on to the population, community, ecosystem, landscape and biospheric levels. In the simplest sense, biodiversity, may be defined as the sum total of species richness, i.e. the number of species of plants, animals and micro-organisms occurring in a given habitat.

Broadly speaking, the term biodiversity includes three different but closely related aspects:

Genetic diversity (diversity within species). It refers to the variation of genes within species. This constitutes distinct population of the same species or genetic variation within population or varieties within a species.

Species diversity (diversity between species). It refers to the variety of species within a region. Such diversity could be measured on the basis of number of species in a region.

Ecosystem diversity. In a ecosystem, there may exist different landforms, each of which supports different and specific vegetation. Ecosystem diversity in contrast to genetic and specific diversity is difficult to measure since the boundaries of the communities which constitute the various sub-ecosystems are elusive. Ecosystem diversity could best be understood if one studies the communities in various ecological niches within the given ecosystem; each community is associated with definite species complexes. These complexes are related to composition and structure of the biodiversity.

Biodiversity at global level

It is estimated that there exists 5-30 million species of living forms on our earth. Of these, only 1.5 million have been identified. These include 3,00,000 species of green plants and fungi, 8,00,000 species of insects, 40,000 species of vertebrates, and 3,60,000 species of micro-organisms (Fig. 13.1).

According to some recent estimates, the number of insects alone may be as high as 10 million, but many believe it is more likely to be around 5 million. Global diversity reserves and country-wise distribution of animal species are shown in Figures 13.2 and 13.3 respectively.

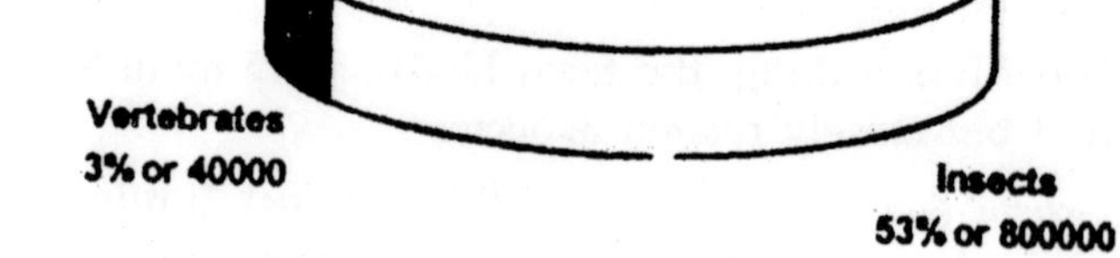

Fig. 13.1. Global distribution of recorded species (1.5 million) of plants, animal and micro-organism.

The tropical forests are regarded as the richest in biodiversity. Scientists are of the opinion that whatever be the absolute numbers, more than half of the species on the earth live in moist tropical forests which is only 7% of the total land surface. Insects (80%) and primates (90%) make up most of the species. For instance, from a single tropical leguminous tree 43 ant species belonging to 26 genera have been retrieved. This approximately equals to the ant diversity of all the British Isles. In 10 selected one hectare plots in Kalimanthan in Indonesia, Peter S. Aston of the Harvard University found more than 700 tree species, almost equal to the number of tree species native to all of North America.

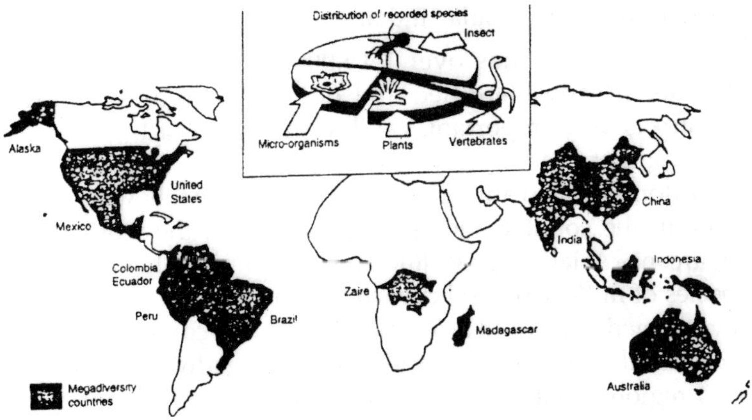

Fig. 13.2. Map of the world showing biodiversity reserves.

The following explanations have been put forward with regard to the high species diversity in tropics:

1. In tropics, conditions for evolution were optimum and for extinction fewer;
2. In tropics, species diversity was conserved over geological time. This is because low rates of extinction prevailing there; and
3. Biological diversity is the result of interaction between climate, organisms, topography, parent soil materials, time and the heredity. The tropics are the ideal place for such an interaction.

However, these explanations need experimental observations and confirmation.

Biodiversity at Country level

The Indian region (8°-30°N and 60°-97.5° E) having a geographical area of 329 million hectares is quite rich in biodiversity with a sizable percentage of endemic flora and fauna. This richness in biodiversity is due to immense variations in climatic and attitudinal conditions coupled with varied ecological habitats. These vary from the humid tropical Western Ghats to the hot desert of Rajasthan, from the cold desert of Ladakh and the icy mountain of Himalayas to the warm costs of peninsular India.

The country has over 1,15,000 species of plants and animals already identified and described. In addition, the country is very important Vavilovian Centre of biodiversity and origin of over 167 important cultivated plant species, and some domesticated animals. To name a few the following crops arose in the country and spread throughout the world: rice, sugarcane, Asiatic vignas, jute, mango, citrus, banana, several species of millets, several cucurbits, some ornamental orchids, several medicinals and aromatics. In fact, our country has been recognised as one of the world's top 12 megadiversity nations. This region is also a secondary centre of diversity for grain amaranthus, maize, red pepper, soyabean, potatoes and rubber plant.

There are several examples of plant germplasm from India making significant contributions to plant improvement. Two cases are well known:

1. Nearly 20 cultivator of rice contain useful genes from wild rices of Kerala, which are responsible for resistance in cultivated rices; and
2. The contribution to the cantaloup (musk melon) industry (14,000 ha) of California by powdery mildew-resistant genes from the wild musk melon of our country.

In flora, the country can boast of 45,000 species which accounts for 15 per cent of the known world plants. The distribution of the recorded flora of the country is shown in Fig. 13.4.

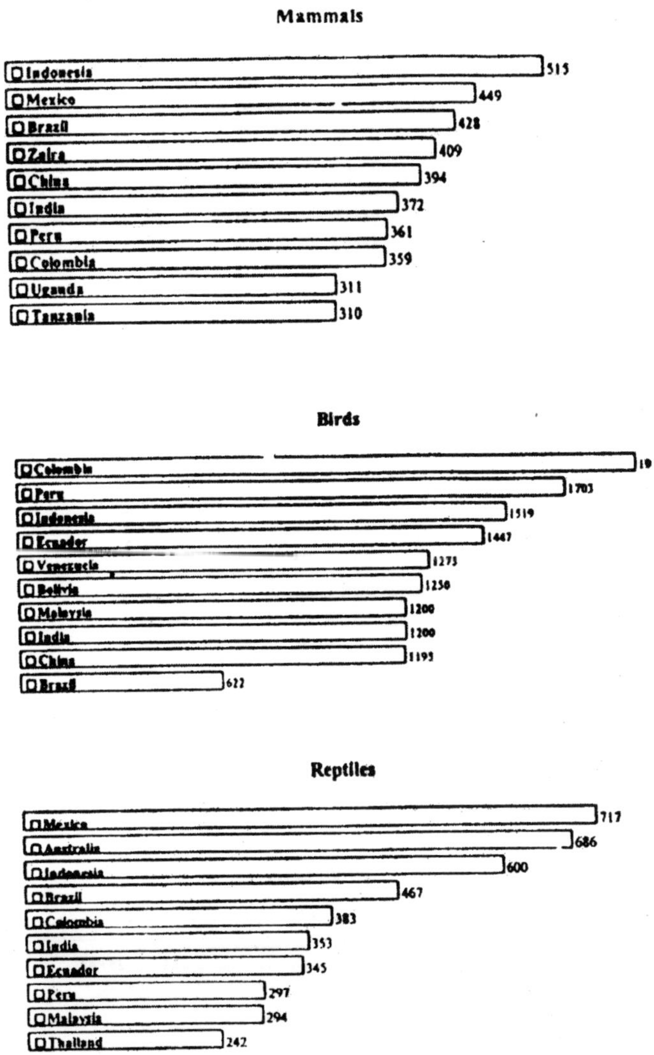

Fig. 13.3. Country-wise distribution of animal species.

Of the 15,000 species of flowering plants, 35 per cent are endemic and located in 26 endemic centres. Among the monocotyledons, out of 588 genera occurring in the country, 22 are strictly endemic.

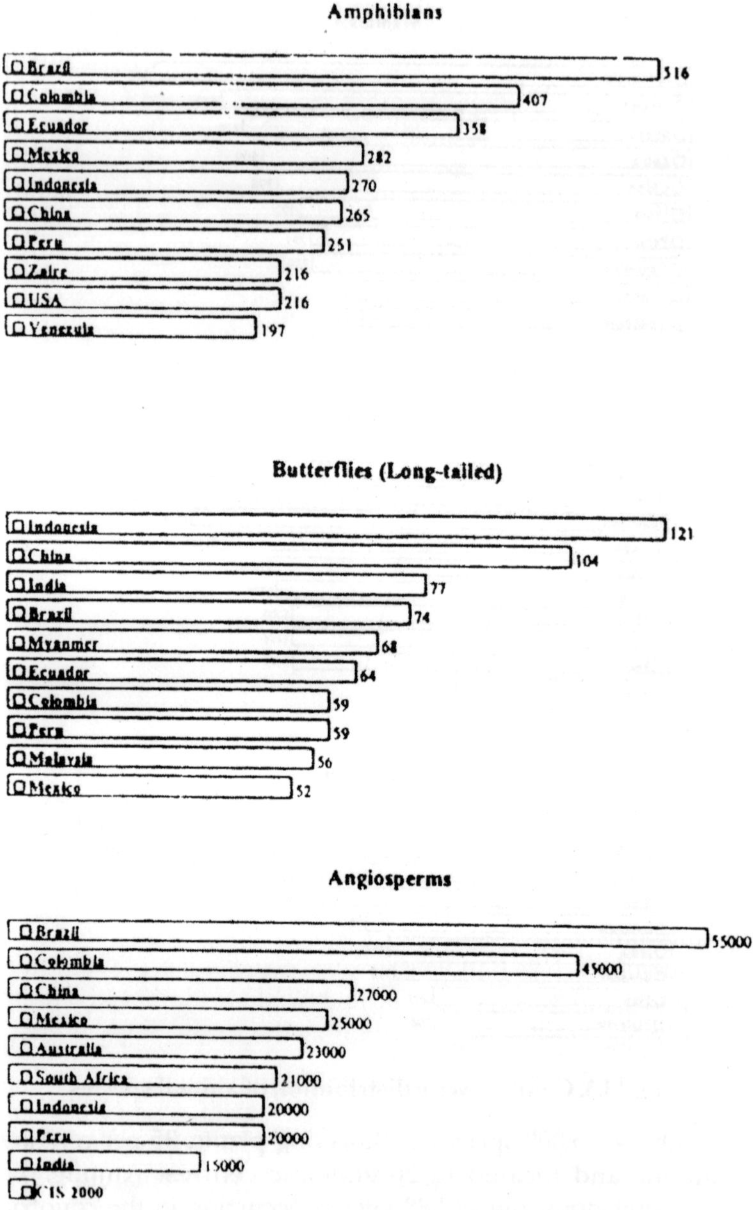

Fig. 13.3. Country-wise distribution of animal species.

The family Poaceae has the highest endemism both by genera and species.

The North Eastern region could boast of being unique treasure house of orchids in the country, the abode of about 675 species out of 1,000 available in the Indian penninsula and against 17,000 species the world over. The important Indian orchids are: *Paphiopedilum fairieyanum (Lmdl) pfitz., Cymbidium aloiflium Sw., Aerides crispum Lindl.,* etc.

Our country is very rich in faunal wealth also. The country has nearly 75,000 animal species, about 80 per cent of which are insects. The distribution of major animal groups are shown in Table 13.0.

Table 13.0. Animal species

Group	*Number of Species*		*Percentage of*	*World*
	World	*India*	*Endemism*	*Percentage*
Mammals	4,231	372	8	8.79
Birds	12,450	1,200	4	9.63
Reptiles	6,300	435	33	6.90
Amphibians	4,184	181	62	4.32
Fishes	23,000	2,000	-	8.69
Insects	8,00,000	60,000	-	7.5
Molluscs	1,00,000	5,000	-	0.5

In animals, the rate of endemism in reptiles is 33% and in amphibians 62%. Further there is wide diversity in domestic animals, such as buffaloes, goats, sheeps, pigs, poultry, horses, camels and yalks. Domesticated animals too have come from the same cradles of civilisation as the major crops.

There are no clear estimates about the marine biota though the coastline is 7,000 km long with a shelf zone of 4,52,460 sq km and extended economic zone of 20,13,410 sq km. There is an abundance of seaweeds, fish, crustaceans, molluscs, corals, reptiles and mammals.

Information regarding other flora and fauna are patchy. Hundreds of new species may be present in our country awaiting discovery. The Western Ghats in Peninsular India, which extend in the southern states are a treasure house of species diversity. Out of the described 15,000 species of the flowering plants in India about 5,000 species occur in the Western Ghats of Kerala; 235 are exclusive to this region. It is estimated that almost one-third of the animal varieties found in India have taken refuge in Western Ghats of Kerala alone.

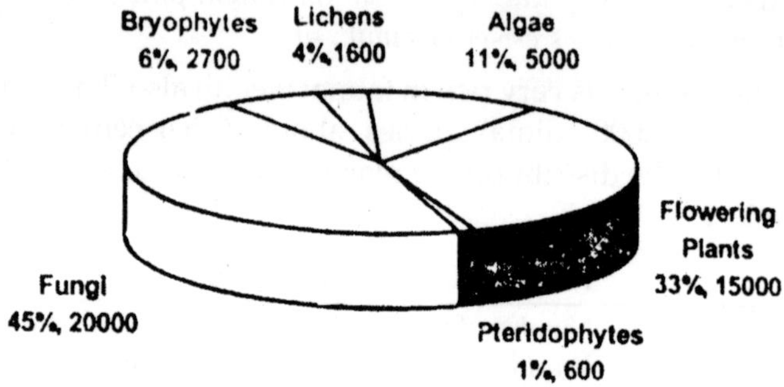

Fig. 13.4. Distribution of recorded (45,000) Indian flora.

Species diversity and ecosystem stability

Species diversity or diversity of genes within species increases its ability to adapt to adverse environmental conditions. When these varieties or populations of these species are destroyed, the genetic diversity within the species is diminished. In many cases, habitat destruction has narrowed the genetic variability of species lowering the ability to adapt to changed environmental conditions. Finally, this diminished genetic variability may lead to ecosystem instability. In other words, the greater the variability of the species, the more is the ecosystem stability.

Ecosystem stability has been considered to be related to the cycling and recycling of nutrients, which in turn, increases the efficiency of the resource use in the ecosystem. The high species diversity thus may be instrumental in cycling and recycling of nutrients and thereby achieving the stability of the ecosystem.

The survival and well being of the present day human population depends on several substances obtained from plants and animals. The nutritional needs of mankind are also met by wild and domesticated animal and plant species. Indeed, the biodiversity in wild and domesticated form, is the source for most of humanity's food, medicine, clothing and housing, much of the cultural diversity, and most of the intellectual and spiritual inspiration. It is, without doubt, the very basis of man's being. It is believed that 1/4th of the known global diversity, which might be useful to making in one way or other, is in serious risk of extinction. This calls for an integrated approach for conserving global biodiversity. Establishment of nature reserves or biospheres with lot of biophysical variability, maintenance of corridors with different nature reserves for the possible migration of the species in response to climate change, etc., are the immediate steps to be taken for conserving the very precious biological diversity on the earth planet. An international consensus on establishing global net-work for gene banks, micro-biological resource centres, and marine parks is also important. At the same time conservation must be coupled with socio-economic development, especially in countries where population pressure threatens the national biotic resources.

14

BIODIVERSITY LOSS

The loss of biological diversity is a global crisis. There is hardly any region on the Earth that is not facing ecological catastrophes. Of the 1.5 million species known to inhabit the Earth (humans are just one of them), one-fourth to one third is likely to extinct within the next few decades. Biological extinction has been a natural phenomenon in geological history. But the rate of extinction was perhaps one species every 1000 years. But man's intervention has speeded up extinction rates all the more. Between 1600 and 1950, the rate of extinction went up to one species every 10 years. Currently, it is perhaps one species every year.

The destruction of the world's tropical forests, which are disappearing at an alarming rate, is one of today's most urgent global environmental issues. A rich species diversity is slowly being lost for ever. Tropical forests are estimated to contain 50 to 90 per cent of the world's biodiversity. According to the report, "People and the Environment", released recently by the US based WRI, the current rate of biodiversity loss is the fastest ever known. The report, based on studies earned out by FAO and WCS, found that the tropical forest is shrinking at the rate of 0.8 per cent each year. Between 1980-90, 154 mha of tropical forests equivalent to almost three times the size of France have been destroyed. Recently, Carl-Dieter

Spranger, the German Minister of Economic Cooperation, pointed out that every year, some 17 mha of tropical forests are destroyed; this is equal to half of the entire area of Germany. If the current rate of deforestation continues scientists estimate that roughly 5 to 10 per cent of tropical forest species may face extinction, within next 30 years. The extent and loss of tropical forests in different ecosystems is shown in Table 14.0.

Table 14.0. Extent and loss of tropical forests in different ecosystems

Forest type	*Extent 1990* ('000 ha)	*Annual decline 1981-90* (Percent)
Rain	713,790	0.6
Moist deciduous	591,779	0.9
Hill and mountain	201,417	1.1
Dry deciduous	178,579	0.9
Very dry	59,742	0.5
Desert	8,086	0.9

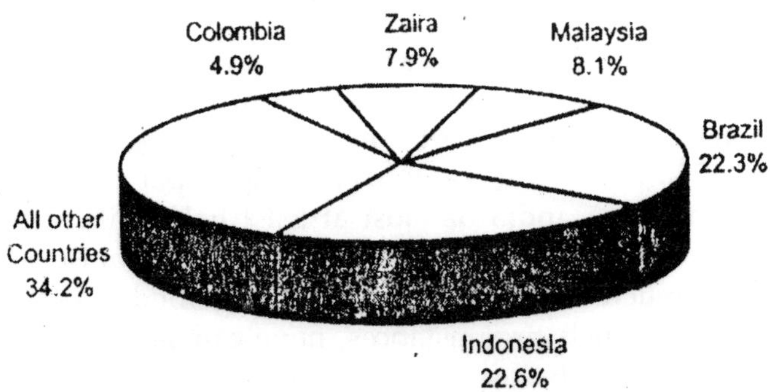

Fig. 14.1. Depicts the share of total tropical rainforests lost during 1981-90.

Rainforest, the home to half of the world's life forms, continue to be destroyed at the rate of over 100,000 Km every year. This loss of biodiversity has immediate and long term effects on human survival. The majority of the world's population still depends on wild plants and animals for their daily food, medicine, housing and household material, agriculture, fodder, fuelwood, spiritual sustenance, and intellectual stimulation.

The loss is even more direct in the case of domesticated biodiversity. Traditional farmers of the world have developed an incredible variety of crops and livestock. This too has been eroded over the last few decades, as lakhs of traditional crop strains and hundreds of domesticated livestock breeds being replaced by a handful of laboratory-generated hybrids or by dominant cash crops.

The traditional diversity was bred to meet diverse human needs of nutrition, test, colour, ritual, smell, and to resist drought, flood and pests. It provided several kinds of insurance against crop failure to the farmer. Modern hybrids, on the other hand, while substantially increasing the grain yield and monetary profits, have forced the farmers to look elsewhere for their other daily needs, especially fodder.

India's biodiversity is one of the most significant in the world. As many as 45,000 species of wild plants and over 77,000 of wild animals have been recorded, which comprise about 6.5 per cent of the world's known wildlife. An assessment of wildlife habitat loss in tropical Asia in 1986 showed that the country had only 6,15,095 Km^2 out of its original wildlife habitat of 30,17,009 Km^2 i.e. loss of about 20 per cent. In the last few decades, India has lost at least half of its forests, polluted over 70 per cent of its water bodies, built on or cultivated much of its grasslands, and degraded most of its coasts. Under such circumstances, none can say how many species have already lost.

The country has several problems such as overpopulation, large number of cattleheads, growing demand for land, energy, and water supply. Unplanned developmental works and over-exploitation of resources have made its living resources most vulnerable. Of the world's 12 top priority biodiversity hot spots, India has two within its boundaries. Over-exploitation has not only resulted in shortages of various materials but also left our biodiversity exposed to various ecological threats. Over emphasis on timber logging has affected many animal species. Faunal losses have been mainly because of over-exploitation of certain species for trading purposes; habitat alteration and destruction; and pollution of streams, lakes and coastal zones.

Listing Of Three Fenei Biodiversity

Red Data Book is the name given to the books dealing with threatened plants or animals of any region. Many countries have prepared their own Red Data Books (e.g. Britain, New Zealand, etc.). On the global level, the IUCN published Red Data Book in two volumes. Its opposite is the Green Book, which lists rare plants growing in protected areas like botanic gardens. A mimeographed Green Book for India has been brought about. It deals with about a hundred rare plant species growing in the garden of Botanical Survey of India (BSI). The BSI has also compiled three volumes of Red Data Book having information on endangered plant species. The UNEP has compiled endangered species of the world under the title, 'Blue Book'. The IUCN has defined Red Data Categories which specify the state of extinction process :

- **Vulnerable:** These are the species whose population numbers are decreasing and are likely to become more severely threatened with time and in near future, they may represent the category of endangered species, if unfavourable conditions in the environment continue to operate.
- **Endangered:** The species with fewer individuals because of unfavourable environmental or human factors and that its natural regeneration is not able to keep pace with exploitation or destruction by natural and unnatural means. If the same factors continue to operate as before the species would extinct soon, e.g., Indian Rhinoceros, Asiatic lion, and the great Indian Bustard.

- **Rare:** The species (or texa) with small world population that are not at present endangered or vulnerable, but are at risk. Such species are usually localised within restricted geographical areas or habitats or are thinly scattered over a more extensive range. Rare species have a population of less than 20,000 individuals. Some species are naturally rare and have never occurred in greater numbers, yet they are able to maintain these numbers. Other species become rare through man's action or other unnatural forces.
- **Extinct:** Species that are no longer known to exist in the wild but survive in cultivation. Generally, the term extinct is used for the species that are no longer known to exist in the wild.
- **Threatened:** It is a broader term that is used for species that fall into any of the above categories.

Threatened Animals

Though IUCN in 1988 listed 23 species of mammals as endangered or vulnerable to extinction, 75 species are totally protected as listed under schedule I of the Wildlife (Protection) Act, 1972. Forest of the Western Ghats is famous for their endemic fauna. The lion-tailed macaques, the Nilgiri Langur, the Malabar large-spotted civet, the Nilgiri tahr, etc., are some important endangered species. The Hoolock gibbon is the only ape found in the hilly forests of north-eastern India. Among the 19 primate species, 12 are endangered. The Manipur brownantlered deer, once distributed in some parts of the north-east, is perhaps the most threatened species. There are some 50 to 100 of them in the Keibul Lamjao sanctuary of Manipur. Over-exploitation is one of the major causes for the present endangered status of the Himalayan musk deer. Human population has also caused depletion in the population of the Kashmir stag or Hangul.

Indian wild ass and buffalo are also among the threatened mammals.

The cheetah, lesser Indian rhinoceros have been extinct according to IUCN's Red Data Book. Other animals appear on the list are: Asiatic lion, snow leopard, swamp deer, elephant and tiger.

According to Dr. G.M. Oza, president of the Indian Society of Naturalists, there are atleast 30 mammals in India classified as threatened and the country tops to the list of countries with endangered mammalian species. After India, the US and China jointly occupy the second place with 29 mammals in both the countries on the threatened species list. Brazil, with 25 endangered species is the third on the list. Further in the oriental region, the country stands first in the threatened mammalian species; Indonesia with 20 threatened species at the second place; Nepal and Thailand, with 14 threatened mammals jointly at the third place. Both Burma and Pakistan have 13 threatened species of mammals and Sri Lanka with only four threatened species.

There are 1175 species of the birds, including 42 endemic, and 14 of these are considered threatened. N.J. Collar and P. Andrew in their book, 'Birds to Watch', have identified 70 species of threatened birds, 63 on the mainland, four on the Andaman Islands and three on the Nicobar Islands. A number of threatened endemic birds are from north-east. The wood-pigeon of the Western Ghats, the white-winged wood-duck of north-east India and the forest owlet of the Satpura hills are among the threatened birds.

IUCN Red data book includes Bengal florican and cheer pheasant as endangered bird species.

The reptile fauna comprises 435 species including 238 species of snakes, about 30 kinds of turtles and many crocodiles and lizards. The tropical forests of Western Ghats alone support some 20 species of snakes. In addition to their commercial exploitation for skin, the dwindling forest ecosystem also exposes these creatures to danger of extinction. The Indian python, a threatened species, is although widely distributed but its population has declined because of habitat destruction.

The four monitor lizards-the common Bengal monitor, the water monitor, the desert monitor and the yellow monitor-have been declared protected. Marine turtles, freshwater tortoises and crocodiles have also been exploited. The green turtle, the loggerhead and the Olive Ridley are the most widely consumed species. All the four species of sea turtles and five of mud turtles have been declared protected. The ghanal, the saltwater crocodile and the marsh crocodile (mugger) are exploited for their valuable skins. The three are included in IUCN list of threatened animals. But the population of crocodiles has been increased through breeding centres in Uttar Pradesh, Andhra Pradesh, Tamil Nadu, Orissa and West Bengal.

The amphibian fauna comprises 182 species, which includes salamanders, caecihans, frogs and toads. There is a high degree of endemism and out of 112 endemics, 84 occur in the Western Ghats and 20 in the north-east. Thereof the Indian species, namely the Himalayan newt, Tylatotriton verrucosus, the Malabar tree-toad, Pedostibes tuberculosus and the Garo Hills tree-toad, Pedostibes kempi; have been identified as rare endangered by B.K. Tikader, in his 'Threatened animals of India'. The first mentioned is the only species of salamander known in the country and are distributed in Sikkim, north West Bengal, Arunachal Pradesh and Manipur. The Malabar tree-toad was recently rediscovered in the Silent Valley.

The status of many invertebrate species is not known. Lack of data is perhaps why Schedule I of the Wildlife (Protection) Act, 1972 includes only a few of the species-the coconut crag of Andaman and Nicobar Islands, the Himalayan dragonfly and 130 species of butterflies. A number of species, beetles and butterflies are also becoming rare or vulnerable to extinct either due to excessive exploitation or habitat destruction. Two important species of butterflies, which are involved in the trade are, Kaiser-i-Hind, the swallow-tail butterfly found along the Eastern Himalayas, and the Bhutan Glory, distributed in Assam, Sikkim, Manipur and Nagaland.

Besides the well-known scred chank of India, Trochus and Turbo are the shell fish that deserve special mention and trade controls.

THREATENED PLANTS

Plants of Ornamental Value

1. *Aerides crispum Lindl.* (orchidaceae) is an orchid with showy flowers; grows in certain regions of peninsular India.
2. *Cymbidium aloiflium Sw.* (Orchidaceae) has not only attractive flowers but is reported to be medicinally useful and hence in great demand.
3. *Diplomeris hirsuta (Orchidaceae).* The species is seriously threatened as it grows in parts of Darjeeling and Meghalaya only and is vulnerable to landslides and forest clearance.
4. *Paphiopedilum fairieyanum (Lindl.)* Pfitz. (Orchidaceae) or 'Lady's Slipper Orchid'. A ground orchid which occurs in Arunachal Pradesh has been badly disturbed due to constant clearance and burning of forests for Jhụm cultivation'. It is so rare that it was once labelled as 'the long lost orchid'.
5. *Qaphiopedilum druna (Orchidaceae).* An orchid of medicinal value found in Kerala, has become very rare.
6. *Rhododendron edgcwortythii* Hk. f. (Ericaceae) is a showy epiphyte, reported to have been collected only on three occasions during the last one hundred years.
7. *Symplocox chengapae* Raiz. & Sahni (Symplocaceae) is an ornamental tree with fragrant flowers, confined to Nicobar Islands.

Plants of Medicinal Value

1. *Acorns calanus* (Araceae) or 'Safed Bach'. The plant species which grew abundantly in Kashmir, Manipur and Nagaland, only a few years back, is now being

imported to meet the medicinal demand of the country.

2. *A tropa acuminata* Royle ex. Lindl. (Solanaceae) or 'Sag-angur'. This species is a source of some active principles. The plant found to grow in Western Himalayas is rare now. Several species of this genus have also become rare.
3. *Dioscorea deltoidca* (Dioscoreaceae). The tubers of the plant species are used to cure rheumatic illness and for opthalmic problems. The species found to grow in the western Himalayas, Punjab and Kashmir is now facing extinction due to over-exploitation.
4. *Drosera species* (Droseraceae). A small insectivorous plant which is not only a botanical curiosity but also famous for its medicinal value. The population of the plant has become rare now due to over-exploitation.
5. *Podophyllum hexandrum* Royle (Podophyllaceae). A Papa or Banbaigan. The species, naturally occurring in north-western Himalayas, has become rare now due to over-exploitation.
6. *Ramvolfia serpentina* Benth. ex. Kurz (Apocyanaceae) or 'Sarpagandha'. The roots of the plant species are highly valuable in medicine. The alkaloid, reserpine prepared from the roots has its value in treating high blood pressure. The species is nearly at the verge of extinction due to over-exploitation for commercial production of reserpine.
7. *Saussurea lappa* Clarke (Asteraceae) or 'Kuth'. Its roots are highly valuable in medicine. The species found to grow in higher altitudes of the Himalayas has become rare.

Plants of Scientific Value

These plants exhibit some morphological interesting features and hence, are of academic interest.

1. *Balanophora involucrata* Hk. f. (balanophoraceae) is a fleshy tuberous parasite on roots of trees, occurring in the Himalaya. It has been collected only on a few occasions so far.
2. *Dischidia benghalensis* Coleb (Asclepiadaceae) is a climber found in eastern Nepal and Sikkim. Leaves develop into pitchers which collect humus and debris and into these, special branched roots develop from the petiole for absorption.
3. *Nepenthes Khasiana* Hk. f. (Nepenthaceae) is an insectivorous pitcher plant. The leaf tips of the plant develop into curious pitcher like structures arched over by lids. The pitchers are up to about 25 cm long. This plant is a special botanical curiosity. The population of this species, restricted to the Khasi hills in Meghalaya, is not enough even to supply educational institutions of our country. The native habitat of the plant is now a gene sanctuary.
4. *Sapria himalayana* Griff. (Hafflesiaceae) occurs as root parasite in Mishmi Hills of Arunachal Pradesh. Its flowers resemble the well-known Rafflesia flower and are up to about 35 cm in diameter.

Plants of Phytogeographic Significance

Most of the Indian endemic plant species show restricted distribution and are rare in every sense needing protection.

1. *Cempegia jainii* Ansari & Kulkarni (Asclepiadaceae) is an erect tuberous herb endemic to K.uiuigin in Maharashtra.
2. *Frcrca inclica* Dalz. (Asclepiadaceae) grows in hilly regions of Maharashtra and has showy flowers.
3. *Glyphochola mysorensis* (Jain & Hem.) Clayton (Poaceae/Grammae) is a grass endemic to Karnataka.
4. *Helicanthes danser* (Lorantahceae) is a genus of parasitic plants, confined to the mountain of peninsular India.

5. *Jainia nicobarica* Balak (Rubiaceae) is a plant endemic to Andaman & Nicobar Islands.
6. *Manisuns divergens* (Hack.) Ktze (Poaceae/Graminae) is a grass probably first collected in peninsular India in 1832. Its second collection was in 1963, after 13 decades. It is confined to Western Ghats.
7. *Willisia warm* (Podostemaceae) is a genus of small herbs which grows on rocks in swift-flowing systems of Western Ghats.

Trees of Forestry Importance

While most trees of sylviculture importance are carefully looked after by the Forest Departments, some rare species do need protection:

1. *Dysoxylum malabaricum* Hedd. (Meliaceae) is a large tree that grows in Western Ghats; its timber is ornamental.
2. *Pterocarpus santalinus* Linn. f. (Fabaceae/Papiolionaceae) or 'Rakta chandan, commonly called Red Sandal, the valuable wavy-grained timber. It yields insoluble blood-red dye, santalin, used for dyeing cloth, leather and wood. It grows in certain localised regions in Andhra Pradesh and its protection and expansion is very necessary.

Other Economic Plants

1. *Decussocarpus wallirhianus* (Presa) de Lauben (Podocarpaceae) is the only confer occurring wild in peninsular India.
2. *Phyllostachys bambusoides* Sieb. & Zucc. (Poaceae/ Grarninae) is common in China and Japan, but reported in India only recently. It is economically important.
3. *Pinus gerardiana* Wall (Pinaceae) or 'Chilgoza'. The fruits of the tree are used as a delicious dry fruit. Its distribution is restricted to a small region in Western Himalaya.

4. *Santalum album* Linn. (Santalaceae) or 'Chandan' occurs commonly but its uncontrolled illicit exploitation and attack by spike disease warrant care and protection to the species.

Causes For The Loss Of Biodiversity

Proximate Causes

The important proximate causes for the loss of biodiversity are as foilows :

1. **Destruction of habitat:** The natural habitat may be destroyed by man for his settlement, grazing grounds, agriculture, mining, industries, highway construction, drainage, dam building, etc. As a consequence of this, the species must either adapt to the changes, move elsewhere or may succumb to predation, starvation or disease and eventually die. In our country, several rare butterfly species are facing extinction with the uncannily swift habitat destruction of the Western Ghats. Of the 370 butterfly species available in the Ghats, up to 70 are at the brink of extinction.

2. **Hunting:** From time immemorial, man has hunted for food. Commercially, wild animals are hunted for their products such as hides and skin, tusk, antlers, fur, meat, pharmaceuticals, perfumes, cosmetics and decoration purposes. For example, in Africa, in recent years 95 per cent of the black rhino populations have exterminated by poachers for their horn. Today, rhino horn fetches more than $ 15,000 in the pharmaceutical market. In the last seven years alone, over one-third Africa's elephants have been wiped out for some 3,000 tonnes of ivory. In the international market, the cost of ivory is about $ 150 per kg, whereas in the Indian market, it varies between Rs. 2,000 to 2,500 per kg. Africa produces about 700 tonnes of raw ivory every year. The major buyers are Japan, Hong Kong, followed by the U.S., Germany and the U.K. The U.S. alone imports about $ 30 million worth of ivory annually. CITES regulations have, to a great extent, reduced illegal trading and poaching of African Tuskers. Carving and sale of African ivory has been banned in India from April 2,

1992. In January, 1987 the Central Government also banned the trade in Indian ivory (Agrawal, 1993).

The scarlet macaw, once common throughout South America, has been eliminated from most of its range in Central America. Several species of spotted cats such as the ocelot and jaguar have been jeopardised by the demand for their fur. In 1962, nearly 70,000 whales were slaughtered. However, international trade in whale products is banned now.

In the country, rhino is hunted for its horns, tiger for bones and skin, musk deer for musk (have medicinal value), elephant for ivory, Gharial and crocodile for their skin, and jackal for thriven fur trade in Kashmir. One of the most publicised commercial hunts is that of whale. The whale-bone or 'baleen' is used to make combs and other products. According to WWF-Report, the spiny tailed lizard of the Indian desert, locally known as Sanda, is being hunted to extract the oil used for making 'Sanda ka tel' for its so called aphrodisiac properties. The lizard is listed in schedule II of the Indian Wildlife Protection Act, 1972, which bans trade in the endangered species. But the WWF says because of the huge demand from the quacks and doctors the killing continue unchecked.

CITES lists nine Indian animal species which have been severely depleted due to international trade. These are Fin Whale (Balenopteru physalus), Himalayan Musk deer (Moschus moschiferus). Green Turtle (Chelonia mydas), Hawksbill Turtle (Eretmochclva imbricata), Olive Ridley Turtle (Dermochelys olivacea), Salt-water Crocodile (Crocodylus porosus), Desert Monitor Lizard (Varanus griseus), Yellow Monitor Lizard (V. flavescens), and Bengal Monitor Lizard (V. bengalensi.i).

Officials of TRAFFIC-India say poaching of the Indian tiger has risen because of the increasing demand from south-east Asian countries and China, where pharmaceutical factories consume the bones of 100 tigers each year. Such demand has decimated the tiger population in China and brought the Russian tiger to the brink of extinction. As a result, in recent years much of the demand has been met by poachers in India. Smuggling of tiger bones and skins is a lucrative business. One kg of tiger bones fetches $ 90 in India and $ 300 in the international market.

Hunting for sport is also a factor for loss of animal biodiversity.

3. **Over-exploitation:** This is one of the main causes of the loss of not only economic species but also biological curiosities like the insectivorous and primitive species and other texa needed for teaching or laboratory work (like Nepenthes, Gnetum, Psilotum, etc.). Commercial exploitation of biodiversity has invariably meant its overuse and eventual destruction. This has been as true in the case of Indian wild mango trees which were turned into plywood as of the whales, that were hunted for tallow, in the oceans.

Plants of medicinal value likc Podophyllurn hexandrum, Coptis teeta, Aconitum, Discorea deltoidea, Rauvolfia serpentina, Qaphiopedilum druri, etc., and horticultural plants like orchids and rhododendrons come under the overexploited category. Faunal losses have been mainly because of over-exploitation. For instance, excessively harvesting of marine organisms such as fish, molluscs, sea-cows and sea-turtles has resulted in extinction of these animals.

4. **Collection for zoo and research:** Animals and plants are collected throughout the world for zoos and biological laboratories for study and research in science and medicine. For example, primates such as monkeys and chimpanzees are sacrificed for research as they have anatomical, genetic and physiological similarities to human beings.

5. **Introduction of exotic species:** Native species are subjected to competition for food and space due to introduction of exotic species for example, introduction of goats and rabbits in the Pacific and Indian regions has resulted in destruction of habitats of several plants, birds and reptiles.

6. **Control of pests and predators:** Predator and pest control measures, generally kill predators that are a

component of balanced ecosystem and may also indiscriminately poison non-target species.

7. **Pollution:** Pollution alters the natural habitat. Water pollution especially injurious to the biotic components of estuary and coastal ecosystem. Toxic wastes entering the water bodies disturb the food chain, and so to the aquatic ecosystems. Insecticides, pesticides, sulphur and nitrogen oxides, acid rain, ozone depletion and global warming too, affect adversely the plant and animal species.

The impact of coastal pollution is also very important. It is seen that coral reefs are being threatened by pollution from industrialisation along the coast, oil transport and offshore mining.

Noise pollution is also the cause of wildlife extinction. This has been evidenced by the Canadian Wildlife Protection Fund. According to a study, Arctic Whales are seen on the verge of extinction as a result of increasing noise of ships, particularly ice breakers and tankers.

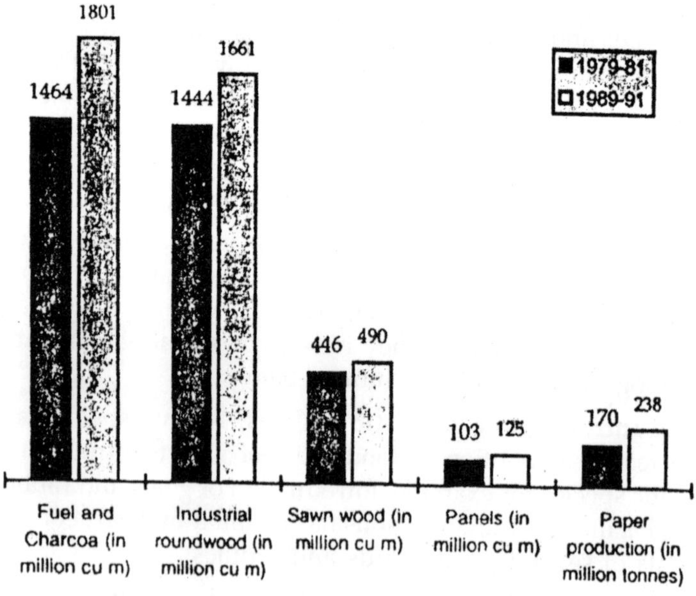

Fig. 14.2. Depicts the increasing production of forest-based materials.

8. **Deforestation:** One of the main causes for the loss of biodiversity is population explosion and resultant deforestation. Deforestation mainly results from population settlement, shifting cultivation, development projects, demand for fuelwood, demand of wood for industry and other commercial purposes. Fig 14.1. depicts the increasing production of forest-based materials. The report, based on studies carried out by the FAO and WCU, found that tropical forest areas are shrinking at the rate of 0.8 per cent each year. Between 1980-90, 154 mha of tropical forests equivalent to almost 3 times the size of France have been converted to other uses. The forest area lost in country is shown in Table 14.1.

Table 14.1. Forest area lost in India due to various human activities during 19511972.

Activity	*Area lost (in ha)*
River-valley projects	4,01,000
Agriculture activities	24,33,000
Roads and communication	55,000
Establishment of Industries	1,25,000
Miscellaneous	3,88,000
Total area lost	**34,02,000**

This represents an average annual loss of 1,55,000 ha.

In India, the rate of deforestation is 13,000 sq. km. annually. If this rate of deforestation continues, one can imagine the ultimate fate of our forest and biological richness. It is presumed that in coming years, the global loss of biodiversity from deforestation alone would be 100 species everyday.

9. **Other factors:** Other ecological factors that may also contribute to the extinction of plant and animal species are as follows

(a) Distribution range. The smaller the range of distribution, the greater the threat of extinction.

(b) Degree of specialisation. The more specialised an organism is, the more vulnerable it is to extinction.

(c) Position of the organism in the food chain. The higher the organism is in food chain, the more susceptible it becomes.

(d) Reproductive rate. Large organisms tend to produce fewer offsprings at widely spaced intervals.

An appropriate break up of human activities that cause threatening of biodiversity is shown in Fig. 14.3.

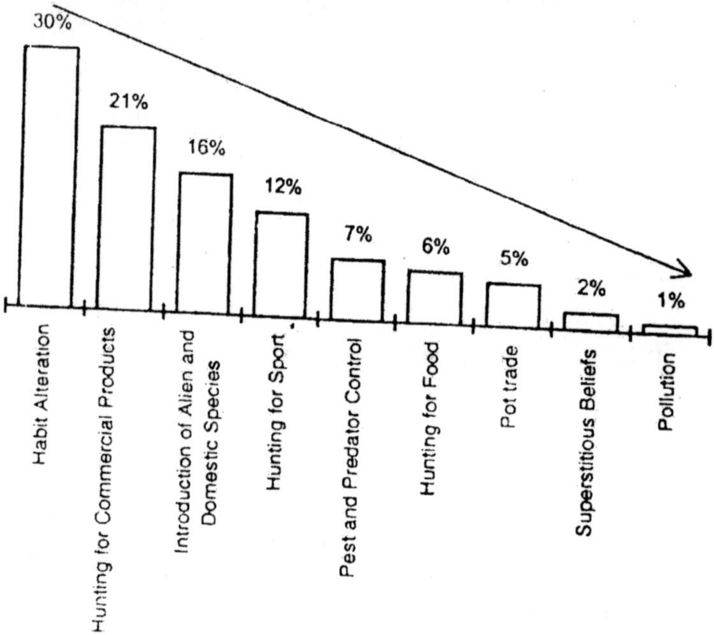

Fig. 14.3. A break up of human activities that cause threatening of biodiversity.

Socio-economic and Political Causes

The socio-economic and political causes of biodiversity loss vary from region to region. In recent times, they can be linked to governmental and international support for industrial forestry, agriculture and energy programmes. The enormous fires in the Amazon have been fueled by two main sources-

State subsidies for the cattle industry (Caufield, 1984; Lutzenberger, 1987); and the taking over of the fertile lands in the North-East and South of Brazil by agri-business operations to grow export crops. As Jose Lutzenberger noted in 1989:

"The promotion of subsidy for big mono-cultures of soyabeans did not make place to feed hungry Brazilians, but to feed fat cows here in the Common Market, cows that are producing what even the technocartx of the European Market call 'seas of milk and mountains of butter'. It was this policy of promoting cash crops in the south that destroyed one of the most well adapted, healthy and ecologically sane peasant cultures that we had in the South of Brazil. Hundreds or thousands of people, may be a million, lost their lands or lost access to land, and these are the people that are now being driven into rainforest".

In South-East Asia, it is the tropical timber industry that is cutting down the last primary rainforests, against the will of local people (Nectoux and Kuroda, 1989; Colchester, 1989; Hong, 1986). As in Brazil, the traditional practices and rights of local communities are violated and the destruction of resources for the benefit of the few is justified in the name of national development and the generation of foreign exchange. The role of western culture, i.e. 'use and throwaway culture' that causes the liquidation of natural resources for short-term profit, is also one of the important causes of deforestation, and thereby the loss of biodiversity in industrialized countries.

The introduction of exotic species of plants and animals causes major disturbances to ecosystems because the checks and balances that hold the equilibrium in the native ecosystem do not apply to the introduced species.

In developing countries like India, the development policies and projects have rarely been sensitive to the need for biodiversity conservation, and that of the local communities. The Government's failure to remove poverty and curb middle class consumerism has led to conditions in which sensible natural resources management assumes low priority.

15

TECHNIQUE OF BIODIVERSITY CONSERVATION & ITS IMPLICATION

Why Conservation?

Indian region is a treasure house of wild genetic resource. Wild species and relatives of crop plants contain valuable genes that are of immense genetic value in crop improvement programmes. The important wild related species of types in various crap groups, prevailing under different phytogeographic zones or the country needs particular attention in the agro-biodiversity management system for a sustainable use to help maintain food, nutritional and agricultural economic security.

It is assumed that these genetic resources shall play a positive and unique role in the development of new cultivars, including restructuring the existing ones that are handicapped due to poor expression or lack of one or the other attribute. The most important characters derived from the wild relatives have been the resistances to biotic and abiotic stresses. Such resistance has been mostly observed to be invariably simply inherited, easily transferable and clearly expressed. Ever since the successful translocation by Scars, of resistance factors in wheat against yellow stripe rust (Puccinia striformis), this approach has been followed in several crops against different diseases, using various techniques and degrees of manipulation. Nevertheless, for certain crops with unique biological cycle and narrow genetic variability in then culiixaied types, such as potato and groundnut breeders and more likely to turn increasingly to the wild relatives of the crops for useful characters.

Conservation of biodiversity through a network of protected areas including National Parks, Sanctuaries, Blaspheme Reserves, Marine Reserves, C.cnc Hanks, Wetlands, Coral Reels, etc.

1. Conservation of micro-organisms, which help in reclamation of wastelands and revival of biological potential of land.
2. Protection of domesticated plant and animal species in order to conserve indigenous genetic diversity.
3. Maintenance of corridors between different nature reserves for the possible migration of species in response to climate, or any other disturbing factor.
4. Rehabilitation of rural poor/tribals displaced due to creation of protected areas.
5. Protection and sustainable use of genetic resources/ germplasm through appropriate laws and practices.
6. By establishing database at various levels to document support for protecting traditional skills and knowledge for conservation.
7. Multiplication and breeding of threatened species through modern techniques of tissue culture and biotechnology.
8. Discouragement of monoculture introduction.
9. Control of over-exploitation through cities and other agencies.
10. Restriction on introduction of exotic species without adequate investigations.

11. To establish conservation parks for the rare, endemic document local resources and support for threatened species.

All these efforts need a change in the attitude of bureaucrats in order to solve the problem, develop harmonious association of the forest departments and forest communities towards being partners in conserving biodiversity, adopting new management approaches to the whole problem, understanding the multiproduct benefits of biodiversity, through incorporating the latest in technical advancement in research and management aspects of forests.

The conservation of biodiversity is linked with the maintenance of ecological stability and productivity. These attributes are important in sustained development and stable national economy (Me Neely, 1988). The concept of biodiversity is very vast and seated at a inicmlevel. In holistic sense, it can be linked with most of the biological processes in nature. Its conservation involves a number of parameter such as number of species, their population dynamics, distribution, habitat, structure, microhabitats, physical environment, climate, present management and past history. Therefore, the efforts for conservation of biodiversity should be in tune with the processes and its occurrence in space and time, from micro-level to megalevel.

Biodiversity is the variety and variations occurring in nature, which has sustained the harmonious existence of life on earth. The components of this diversity are so interdependent that any change in the system leads to a major unbalance and threatens the normal ecological cycle.

Acknowledging the need for conservation, the concern for conservation of biodiversity at global level figured for the first time. In the discussions at the UN Conference on the Human Environment held at Stockholm in 1972. The UNEP identified conservation as a priority area in 1972. It was only towards 1980's that systematic and concentrated efforts to look at biodiversity conservation profile at international level started with constitution of an ad hoc working group of experts on biological diversity by UNEP in 1987. Eventually, an expert group was constituted by UNEP which started it's work in 1989.

Culmination in Conversion on Biological Diversity [CBD] at the UN Conference on Environment and Development [UNCED] held at Oleo de Janeiro in June 1992. This convention entered into force on 29th December, 1993. At present, 166 countries are parties to the convention.

This international treaty is a historic treaty in that it not only reflects the commitment of global community for conservation and sustainable use of biodiversity but also visualises sharing of benefits arising out of utilisation of genetic resources with the countries of origin.

Objectives

According to this, the main objectives of the convention arc:

1. The conservation of biological diversity.
2. The sustainable use of components of biodiversity.
3. The fair and equitable sharing of benefits arising out of the utilisation of genetic resources.

The objective of equitable sharing of benefits has to be further seen in the reaffirmation of sovereign rights of states over their own biodiversity which entitles a country to equity and fairness in sharing of benefits arising out of the utilisation of their genetic resources. The 23 perambular paragraphs of the convention recognise and reaffirm.

1. The intrinsic value of biological diversity.
2. The sovereign rights of states over their biological resources.
3. The fundamental requirement of *in-situ* conservation of ecosystem and natural habitats.
4. The supporting role of *ex-situ* measure.
5. The vital role of local communities and women in the conservation and sustainable use of biological diversity.

6. The desirability of sharing equitably the benefits arising from the use of traditional knowledge, skits, innovations and practices.
7. The importance of and the need to promote regional and global cooperation for conservation, and
8. The requirement of substantial investments to conserve biological diversity.

Implications

The significant implications of the provisions of the convention and the main obligations imposed on the contracting parties are summarized below:

1. The parties are required to take measure for *in-situ* conservation of biological diversity, promote rehabilitation and restoration of degraded ecosystem, and ensure protection of threatened species (Article 8).
2. The parties are obliged to respect, preserve and maintain knowledge, innovation and practices (Article 8).
3. The parties are also to adopt measures for *ex-situ* conservation of components of biological diversity, for complementing the *in-situ* efforts (Article 9).
4. The parties are to facilitate access to genetic resources on mutually agreed upon terms with prior informed consent of the country providing these resources (Article 15).
5. The recipient country is to share in a fair and equitable way, the results of the research and development and the benefits arising from the commercial and other utilisation of genetic resources with the party providing such resources. The convention calls for transfer of relevant technologies, including biotechnology on fair and most favourable terms from the developed to developing nations which provide genetic resources used for the development of these technologies (Article 16).

6. It also calls on the private sector to facilitate access to and transfer of such technologies developed by them (Article 16.4).
7. The contracting parties are to cooperate in this regard to ensure that patents and other intellectual property rights are supportive of and do not run counter to the objectives of the convention (Article 16.5).
8. Recombinant DNA technology (or genetic engineering) are increasingly gaining ground and are being seen as potentially useful for various sectors by experts. Recognizing that introduction of these biotechnologies requires utmost care and caution based on precautionary principal (outlines in Agenda 21) particularly because recall of an introduced organism in nature would well be possible, the convention commits the parties to consider an international protocol for safe transfer, handling and use of any living modified organism resulting from biotechnology (Article 19.3).
9. The parties are also to take measure for facilitation access on a fair and equitable basis and on mutually agreed upon terms to the results and benefits arising from biotechnologies based upon genetic resources provide/transferred (Article 19.2).
10. The developed country parties are committed to contribute to a fund to enable developing country parties to meet the "agreed full incremental cost' for implementing the provisions for their convention (Article 20.2).
11. The financial mechanism is to operate within a democratic and transparent system of governance and "function under the authority" of the conference of the parties (Article 21).

Important points to note about convention are the faciliation of access by developing countries is linked with equitable sharing of benefits; thus making it a two-way process. Technology News are also found on the principle of equity with the developing countries providing resources and traditional technologies, and developed countries sharing and transferring technologies including biotechnologies and providing financial resources to help developing countries to meet their commitments and realize benefits.

From the foregoing discussion, it is amply clear that the convention is the first global, comprehensive agreement to address all aspects of biological diversity, genetic resources, species and ecosystems which would have revolutionary and far reaching implications. Being based on considerations of equity and shared responsibility, the convention envisages reciprocity of arrangements between developed and developing countries, thereby promoting a renewed partnership between them.

The main implementation measures for the convention are to be through national strategies, plan or programmes, to be developed in accordance with each country's situation and capabilities. Although in its text, the convention commits the contracting parties to take substantive action in many areas as under :

Action Plans

1. Development of National plans, strategies or programmes for conservation and sustainable use of biodiversity and integrating these into relevant sectoral or cross-sectoral plants, programmes and policies.
2. Inventorisation and monitoring of components of biodiversity and of processes adversely impacting it; developing and strengthening of *in-situ* mechanisms for biodiversity conservation both within and outside protected areas.
3. Development of *ex situ* measures for biodiversity conservation, as a complement to *in situ* approaches.
4. Restoration of degraded ecosystem and recovery of endangered species.

5. Adopting measures to avoid and minimize advise impacts on biodiversity.
6. Protecting and encouraging customary use of biological resources that are compatible with conservation or sustainable use of requirements.
7. Adopting economically and socially sound measures that act as incentives for conservation and sustainable use of components of biodiversity.
8. Promoting and encouraging research contributing towards achieving the objectives of the convention.
9. Developing educational and public awareness programmes with respect to conservation and sustainable use of biodiversity.
10. Facilitating access to genetic resources on mutually agreed terms and prior informed consent, and taking measures for fair and equitable sharing of benefits arising from utilisation of the resources thus transferred.
11. Facilitating access to and transfer of technology, including biotechnology to developing countries under fair and most favourable terms.
12. Facilitating exchange of information relevant to biodiversity.
13. Promoting scientific and technical co-operation with other parties.
14. Consideration of a protocol for safe transfer, handling and use of living modified organisms resulting from biotechnology and
15. Providing new and additional financial resources by the developed country parties to enable the developing country parties to meet the agreed full incremental costs for implementing the provisions of this convention.

Conservation For Sustainable Use: A Holistic Approach

In view of the large scale exploitation of useful diversity of various plant species from forest and open areas, particularly for medicinal and other economic plants, and the prevalence of fragile ecosystems in many parts of the country and also the existence of diversity in several useful co-existing biological species, it becomes important to conserve these plant species/coexisting species and systems either by way of domestication and cultivation or by other *ex situ* or *in situ* conservation methods for their sustainable use. The biological factors that determine the conservation approaches and the most common methods used for germplasm conservation of different IGR categories, were clearly defined viz., ecosystems/agro-ecosystems, specific habitats, gene pools, special genetic stocks, etc. A balanced application of these, *ex situ*/on farm/*in-situ*/in vitro conservation approaches and technologies, would be needed so as to suitably register location specificity, case-to-case basis, costeffectiveness, feasibility for retrieval and networking, etc. There is a need to give emphasis on 'cultivation of wild forms' (domestication), rather than 'collecting from the wild alone' because the former lays more emphasis, over the latter approach, on botanical identity, quantity for supply, genetic improvement, quality control, etc., of the conserved material. Such cultivation (domestication) may have to be initiated in the respective agro-habitats showing micro-climates similar to the niche of wild species.

Conservation strategies must have a holistic approach and also dynamism and vibrance of genetic evolution that should be adopted for sustainable utilization of natural resources linked equitable sharing of benefits accrued from the use of plant-biological diversity of the Indian origin.

A collaborative approach should be followed by the national agencies and regional bodies responsible for policy, planning, research and development for agriculture (including horticulture) and forestry to cover agro-biodiversity and biodiversity (wild species) simultaneously.

Diversity mapping should be done along with inventorization at micro-level for effective management, conservation and use of crop plant species and their relatives.

Priorities drawn for future survey, collection and inventorization of PGR diversity should be based on gaps identified.

Agro-ecological niches should be clearly identified along with the prevalent crops and cropping systems, the custodians of diversity, i.e., indigenous community(ies), then ethnic groups and economic background and other developmental and ecological factors related to the area for consideration of their suitability for on-farm conservation may be demarcated for action-oriented programme.

Sub-regional consortia and regional networking should enable to safeguard the regional interests for genetic diversity. Contiguity of adjoining areas housing similar diversity for particular species genepool can be availed, to share the benefits of on farm conservation/*In-situ* conservation across the geo-political barriers.

Ample research back up should be developed, ensuring funds and other resources, at the national gene bank to enable effective scientific monitoring, guidance and future direction to such activities.

Linkages should be strengthened with other governmental departments, state universities, non-governmental organizations and communities and gradually the role of each interested party should be clearly defined for getting better results.

Specially designated genetic reserves, gene sanctuaries and/or genetic gardens should be earmarked for wild species of food value or other economic importance. Conversely, such species should be carefully monitored and inventoried in the existing biosphere reserves. Strategies, approaches and technology, including biotechnology, should be developed/standardized for gradually bringing these wild economic species under cultivation.

Degraded Ecosystem And Conservation Of Biodiversity

Biodiversity is a very fast term that includes almost all living organisms existing on earth. According to Jain (1994), all the great variety that exists in living organisms, plants, animals and microbes on earth reflects biodiversity. The need for conservation of biodiversity significantly increases in degraded and fragile ecosystems to restore productivity, and to arrest further degradation of such areas. According to a survey conducted by the National Wasteland Development Board, New Delhi, out of 329 million ha of the total geographic area of India, about 158 million ha come under different kind of land degradation. Hence, about 50% land in our country constitutes the degraded and fragile ecosystems. Such type of degradation occurs due to several reasons like soil salinity and alkalinity, ravines, brackish water and line to certain other biotic and climatic factors. An only salt affected soil constitutes 7.5 mha area in different part of country. Destruction of vegetative cover, waterlogging due to seepage, and lack of proper drainage are the main causes for the formation of such type of degraded land.

The ideal system for genetic resources conservation is *in situ*, i.e. in reserved forests and protected areas. These include National Parks, Sanctuaries, Nature Reserves, Biosphere Reserves and several others. These areas vary considerably in size, design, purpose and effectiveness of management. The country has nearly 500 protected areas covering 4.6 per cent of the geographical area. However, biodiversity is declining even in these protected areas owing to tremendous and ever growing human and cattle pressure.

Ex-situ conservation, which is conservation outside natural habitats, has an important role to play. This is done by perpetuating sample populations in genetic resource centres, zoos, botanical gardens, culture collections, seed banks, sperm banks, etc., or in the form of germplasm bank in vitro. In India, CRISAT, National Bureau Plant Genetic Resources, Indian Council of Agricultural Research, and a number of other Research Centres are collecting and preserving germplasm.

of a number of plant and animal species. Cryopreservation is superior in this respect. In vitro regeneration of cryopreserved plant species have been made possible in both crop as well as tree species. In animals also cryopreserved eggs and sperms have been fertilized successfully.

However, only conservation of germplasm under laboratory conditions will not solve the problem of maintaining the requisite green cover for ecological equilibrium. In this context the Chinese perception is worth mentioning.

Although, the Government has introduced environmental education as a component, both at formal (School, College and University) and non-formal (Adult education, activities for children, ecodevelopment Camps, training of Senior Executives/Administrators and foundation courses) levels, these are mainly targeting the urban and semi-urban people. But the major concern about maintenance of protected and reserved forests is the pressures of tribals and forest dwellers. It is obvious that the dependence of tribals on forests is maximum and their long-term interest lies in protection and not in destroying forests. Therefore, the tribals and forest dwelling communities must be taken into confidence and should be involved in protecting and conserving the forest heritage. To elicit their cooperation, both Governmental and Non-Governmental Organisations, should be involved.

In this context, it is necessary to know and identify the details of the customs, cultural heritage and nature of dependency of the tribals on forests. This should be worked out independently for different areas and individual tribal groups, as each would represent a unique case.

Although, Article 48A and 51 A(g) clearly states the role and responsibility of the Government and every citizen in protecting and improving the natural environments, there is no provision in the constitution for enforcing them. It is felt necessary to formulate a rule for effective controlled grazing, as grazing leads to the erosion of natural diversity and the area is taken over by certain dominant plant species, which are mostly weeds.

India's Efforts For Biodiversity Conservation

Dr. M. S. Swaminathan (1983), reviewed the scientific aspects of conservation. He suggested that the first step in conservation should be defining the categories of materials (plants/genes) for preservation and the major methods for preserving them. He suggested that the following categories should be usually regarded as important :

1. Cultivated varieties in current use.
2. Obsolete cultivars.
3. Primitive cultivars or land races.
4. Wild species and weedy species closely related to cultivated varieties.
5. Wild species of potential value to man.
6. Special genetic stock developed by man.

The Principle of any technology designed for germplasm conservation should be to preserve the maximum possible genetic diversity of a particular plant or genetic stock for future use. Diversity in plants is at the level of species, varieties and individuals. Special genetic stocks may include the material (mutant or breed lines with identified gene or gene combination) developed and used in ongoing breeding-programme. It has been estimated that the survival of approximately 9,000 wild species of plants is some way threatened and that the majority of these are from tropical regions. This further highlights the need for a positive approach to conservation of endangered plants. When new cultivars replace the primitive or conventionally used agricultural crops, it becomes especially important that the crop be properly documented and conserved.

According to Swaminathan (1983), conservation methodologies can take the following three forms:

1. Entire biomass
2. *In situ* preservation
3. *Ex situ* preservation

Many ways are being suggested for preserving biodiversity. Some of them include following:

1. No undisturbed land use.
2. Catalogues of genetic resources and national biological inventories be prepared.
3. Measures should be taken to reduce the emission of green house gases and ozone destroying compounds.
4. Effective measures for the conservation of biodiversity be developed and strengthened in all countries.

The details of these conservation methods make a separate subdiscipline of biodiversity studies called conservation biology. These techniques of conservation of biodiversity include:

1. *In situ* conservation.
2. *Ex situ* conservation.

Conservation of biodiversity can be achieved in a number of complementary ways. These methods, all of them falling within the broader concepts of gene banks.

In situ Conservation

In situ means in the natural, original place or position; as in the location of the explain on the mother plant prior to excision. *In situ* conservation which include conservation of plant and animals in their native ecosystems or even in man-made ecosystem, where they naturally occur. This type of conservation applies only to wild fauna and flora and not to the domesticated animals and plants, because conservation is achieved by protection of populations in nature. This method of conservation mainly aims at preservation of land races with wild relatives in which genetic exists and/or in which the weedy/wild forms present hybrids with related cultivars. These are evolutionary systems that are difficult for plant breeders to stimulate and should not be knowingly destroyed.

The *in situ* conservation of habitats has received high priority in the world conservation strategy programmes launched since 1980. Institutional, arrangement, especially in countries of the developing world, have been emphasized. This mode of conservation has some limitations however; there is risk of material being lost due to environmental hazards. Further the cost of a maintaining a large portion of available genotypes in nurseries or fields may be extremely high.

In-situ conservation includes a system of protected areas of different categories, e.g., National parks, Sanctuaries, National Monument, Cultural landscapes. Biosphere Reserves etc. One of the best methods to save wildlife species, which is on the road to extinction, is to put it in a special enclosure to reproduce. Sanctuaries and National Parks, whose legal definition varies from country to country, best illustrate this.

National Parks, Wildlife Sanctuaries and Biosphere Reserves

National Parks or a Sanctuary may be defined as an area, declared by stale, for the purpose of protecting, propagating or developing wild life therein, or its natural environment for their scientific, educational and recreational value

The creation of National Parks, Wildlife Sanctuaries and Biosphere Reserves is an attempt to manage wildlife by defining protected areas. Wildlife therein is regularly monitored and necessary management strategies for their perpetuation and preservation are formulated and implemented. These protected areas not only benefit wildlife, but also indirectly humans too. Their protection means the protection of entire ecosystem, which is necessary to continue to enjoy the benefits that we may now receive from it.

A National Park is an area dedicated to conserve the scenery (or environment) and natural objects and the wildlife therein. In national parks, all private rights are non-existent and all forestry operations and other usages such as grazing of domestic animals are prohibited. However, the general public may enter it for the purpose of observation and study. Certain parts of the park are developed for tourism in such a way that enjoyment will not disturb or scare the animals.

Table 15.0. Important National Parks and Wildlife Sanctuaries in India

States	*National Parks and Wildlife Sanctuaries*
Andhra Pradesh	Pakhal Wildlife Sanctuary, Pocharam Wildlife Sanctuary, Kawal Wildlife Sanctuary, Kolleru Pelicanary
Arunachal Pradesh	Namidapha Wildlife Sanctuary*
Assam	Kaziranga National Park*, Manas Wildlife Sanctuary*
Bihar	Hazaribagh National Park, Bella National Park
Goa	Mollen Wildlife Sanctuary
Gujrat	Gir National Park, Velavadar National Park, Wild Ass Sanctuary, Nal Sarovar Bird Sanctuary
Haryana	Sultanpur Lake Bird Sanctuary
Himachal Pradesh	Sechu-tun-Nallah Sanctuary
Jammu & Kashmir	Dechigam Wildlife Sanctuary
Karnataka	Bandipur National Park*, Nagarhole National Park, Ranganthitlo Bird Sanctuary, Silent Valley National Park
Kerala	Periyar Wildlife Sanctuary*, Wynad Wildlife Sanctuary
Madhya Pradesh	Kanha National Park*, Shivpuri National Park, Bandhavgarh National Park*, Parma National Park
Maharashtra	Tadoda National Park, Yawal Wildlife Sanctuary
Manipur	Keibul Lamiao National Park
Meghalaya	Baipakram Sanctuary
Mizoram	Dampa Wildlife Sanctuary
Nagaland	Intangki Wildlife Sanctuary

Orissa	Simlipal National Park*, Chilka Lake Bird Sanctuary
Punjab	Abohar Wildlife Sanctuary
Rajasthan	Ranthambore National Park*, Sariska Wildlife Sanctuary*, Ghana Bird Sanctuary
Sikkim	KaiiGhenjuga National Park
Tamil Nadu	Mudumalai Wildlife Sanctuary, Vedanthangal Water Bird Sanctuary
Uttar Pradesh	Corbet National Park*, Rajaji National Park, Dudhwa National Park*
West Bengal	Jaldapara Wildlife Sanctuary

*Taken Under Project Tiger

Table 15.1. National Parks and Wildlife Sanctuaries

State And Union Territory	*Total Number*	*Sanctuaries Area (Sq. Km.)*	*Total Number*	*National Park Area (Sq. Km.)*
Andaman and Nicobar Islands	94	372.15	8	900.77
Andhra Pradesh	21	11837.54	4	372.23
Arunachal Pradesh	09	6177.45	2	2468.23
Assam	08	990.58	2	930.0
Bihar	19	3881.75	2	567.32
Goa	04	355.78	1	107.00
Gujarat	21	16970.16	4	479.67
Haryana	10	342.65	1	1.43
Himachal Pradesh	30	40702.87	2	1295.00
Jammu & Kashmir	15	10157.67	4	3900.07
Karnataka	20	4238.21	5	2471.98
Kerala	12	2143.36	3	536.52
Madhya Pradesh	32	10567.05	1	6485.72
Maharashtra	25	13995.49	5	958.45
Manipur	01	184.85	2	81.00
Meghalaya	03	34.21	2	267.48

Mizoram	03	560.00	2	250.00
Nagaland	03	24.41	1	202.02
Orissa	18	6214.96	2	1212.70
Punjab	06	249.82	0	-
Rajasthan	06	249.82	4	3856.53
Sikkim	04	92.10	1	850.00
Tamil Nadu	17	2671.03	5	401.63
Tripura	04	603.62	0	-
Uttar Pradesh	29	8107.52	7	5429.83
West Bengal	15	1055.52	5	1695.65
Daman and Diu	01	2.18	-	-
Delhi	01	13.20		
Chandigarh	01	25.42		

The definition adopted by IUCN (1975) is as follows:

National Park

A national park is a relatively large area where-

1. One or several ecosystems are not materially altered by human exploitation and occupation, where plant and animal species, geomorphological sites and habitats are of special scientific, educative and recreative interest or which contain a natural landscape of great beauty.
2. The highest competent authority of the country has taken steps to prevent or eliminate as soon as possible exploitation or occupation in the whole area and to enforce effectively the respect of ecological, geomorphological or aesthetic features which have led to its establishment.
3. Visitors are allowed to enter, under special conditions, for inspirational, cultural and recreative purposes.

Wildlife Sanctuary

A wildlife sanctuary, similar to national park, is dedicated to protect the wildlife, but it considers the conservation of species only and also the boundary of it is not limited by state legislation. Further, in the sanctuary, killing, hunting or capturing of any species of birds and mammals is prohibited except by or under the control of highest authority in the department responsible for management of the sanctuary. Private ownership may be allowed to continue in a sanctuary, and forestry and other usages permitted to the extent that they do not adversely affect wildlife.

The number of national parks and sanctuaries has increased from 33 in 1952 to a total of 221 by the end of December 1980 covering 2-3% of the total geographical area and 10% of the total forest area of the country. Today, India has 75 National Parks and 421 Wildlife Sanctuaries covering an area of about 1.4 lakh km^2-constituting more than 4% of the total geographic area of the country, and one-fifth of all forest areas. The protected area also include 20 tiger reserves and 14 biosphere reserves.

The difference between a Sanctuary and a National Park is subtle and even confusing. Hunting without permit is prohibited and grazing or movement of cattle is regulated in Sanctuary. But hunting and grazing are absolutely prohibited in National Park which may be established within or outside a sanctuary i.e. the difference is mostly the interference of human or human activities in the area. In a sanctuary, the human activities are allowed but in a National Park the human interference is totally prohibited.

Biosphere Reserves

Biosphere reserves aim at conserving the biological diversity and genetic integrity of plants, animals and micro-organisms in their totality as part of the natural ecosystems, so as to ensure their self perpetuation and unhindered evolution of the living resources.

Biosphere reserves have been described as undisturbed natural areas for scientific study as well as areas in which conditions of disturbance are under control. They have been set aside for ecological research and habitat preservation.

Ramde (1984) described them as the means to protect ecosystems, whether natural or modified by human activity, in order to preserve ecological 'evidence' for the purpose of scientific research. Indeed, biosphere reserves are very good methods for implementation of the WCS, 1980.

Zones of Biosphere Reserve

1. Core zone forming the sanctum sanctorum. The core zone invariably encompasses watershed areas, along with specific habitats to be conserved in its original form and therefore to be protected from even non-native or exotic plants and animals. The area may possibly be open strictly to supervise entry by scientists, research workers and conservation authorities.
2. Buffer zone that concentrically surrounds the core zone. Buffer zone bears the necessary as well as inevitable human stress and also meets the research, education and where possible aesthetic needs of society too. In this zone controlled exploitation of natural resources is possible.

Besides these two zones there could be some more zones depending upon the geographical, ecological and cultural situations, such as Forestry zone, Tourism zone, Agriculture zone and Restoration or reclamation zone. Sometimes, two or more core areas are to be protected forming a cluster type of biosphere reserve.

Biosphere Reserve Network Programme was launched by UNESCO in 1971 under as MAB. The objectives of the programme were:

1. Conserve biotic diversity for ecological evidence.
2. Safeguard genetic diversity for the process of evolution to act upon.
3. Provide natural areas for basic and applied research in ecology and environmental biology.
4. Provide opportunity for environmental education and training.

5. Promote international cooperation.
6. Promote appropriate sustainable management of the available biotic resources.
7. Disseminate the experience so as to promote sustainable development elsewhere.

Biosphere Reserves in India

Table 15.2. Biosphere Reserves

Biosphere Reserves	*State(s)*
Nilgiri	Tamil Nadu, Kerala and Karnataka
Namdapha	Arunachal Pradesh
Nanda Devi	Uttar Pradesh
Uttarakhand (Valley of Flowers)	Uttar Pradesh
North Islands of Andamans	Andamans & Nicobar
Gulf of Mannar	Tamil Nadu
Kaziranga	Assam
Sunderbans	West Bengal
Thar Desert	Rajasthan
Marmas	Assam
Kanha	Madhya Pradesh
Nokrek (Tura range)	Meghalaya
Little Rann of Kutch	Gujarat
Great Nicobar Island	Andamans & Nicobar

Table 15.3. List of Biosphere Reserves

Name of Biosphere Reserve	*State/union territory*	*Area in sq. km.*	*Data of Set-up*
Nanda Devi	Uttar Pradesh	2,236.74	18.01.1988
Nokrek	Meghalaya	80.00	01.09.1988
Nokrek, Manas	Assam	600.00	14.03.1989

Sunderbans	West Bengal	9,636.00	29.03.1989
Gulf of Mannar	Tamil Nadu	10,500.00	18.02.1989
Nilgiri	Karnataka, Kerala & Tamil Nadu	5,520.00	01.08.1986
Great Nicobar Islands	Andaman & Nicobar	885.00	06.01.1989
Similipal	Orissa	2,750.00	21.06.1994

Desert Biosphere Reserve yet to be established.

A protected area that can be declared as a biosphere, should satisfy the following essential criteria:

1. Should pinvule a network of projected terrestrial and coastal environments which form a coherent system on a world scale.
2. Should occur in each of the 193 biogeographical provinces of the world distinguished in the classification of Udvardy (1975), so as to exhibit the maximum genetic diversity.
3. Should show a complete range of the different types of human interference, from ecosystems untouched to those which have been degraded by humans.
4. Structure and size should ensure the efficient conservation of the desired ecosystems.
5. Should have sufficient resources available for ecological education, training and research to be carried on in respect to conservation of nature.
6. If possible, should have geographical continuity with other types of protected areas.
7. Should have adequate long-term legal protection.

As of January 1989, 274 biosphere reserves have been established in 68 countries. The list includes the 14 sites proposed as biosphere reserves in India.

In India, the National MAB Committee in 1973, operating under the aegis of the Department of Environment has identified 13 potential sites for designation as biosphere reserves. The fourteenth site, the Great Nicobar Island was added to the list in 1989. The country's first biosphere reserve came into being in 1986 in Nilgiri, covering 5520 km^2 in Tamil Nadu, Kerala and Karnataka. The second the Nanda Devi in Uttar Pradesh and the third the Nokrek in Meghalaya were designated in 1988. The fourth, the Gulf of Mannar in Tamil Nadu was declared in February 1989 and the fifth, to include Marmas in Assam on March 14, 1989. Sunderbans in West Bengal has recently been earmarked as a biosphere reserve. Other areas for future designation are Namdapha, Uttara-khand Kaziranga, Thar Desert, Kanha, Little Kami of Kutch, and North Islands of Andamans.

From the biospheric point of view the region falls into two realms and 12 biogeographical provinces: Ladakh, Himalayan highlands and Malabar rain forest, Bengal rain forest, Indus-Ganga monsoon forest, Assam-Burma monsoon forest, Mahanadian, Coromandel Deccan thorn forest, Thar Desert, Lakshadweep Islands, and Andaman and Nicobar Islands. The first two provinces belong to the Palaearctic realm while the remaining to the Indo-Malayan (Khoshoo, 1991).

Essential Criteria of Biosphere Reserves

1. Represent an ecological protectorate.
2. Occur in a definite biogeographic province (biosphere reserves cover 9 out of the 12 biogeographical provinces).
3. Contain abundant genetic diversity (India harbours nearly 45,000 plant and 65,000 animal species).
4. Have complete range of human interference.
5. Have structure and size sufficient to ensure efficient conservation.
6. Have many opportunities for research in ecology/ environment, population, genetics, evolutionary biology, plant-animal interaction, eco-development, etc.
7. Receive adequate long-term legal protection.

Current Status of Protected Areas

Approximately 4.2% of the total geographical area of the country has been earmarked for extensive *in situ* conservation habitats and a protected area network of 8 Biosphere Reserves, 85 National Parks and 448 Wildlife Sanctuaries has been created. The results of this network have been significant in restoring viable populations of large mammals such as tiger, lion, rhinoceros, crocodiles, elephants, etc. Following are the State-wise data of protected areas.

To conserve the representative ecosystem a Biosphere Reserve Programme has been implemented. Eight biodiversity rich area of the country have been designated has Biosphere Reserve applying the UNESCO/MAB criteria. Details on these Biosphere Reserves are tabulated. These reserves aim at conserving the biological diversity and genetic integrity of plants, animals and micro-organisms in their totality as part of the natural ecosystems, so as to ensure their self-perpetuation and unhindered evolution of the living resources.

The Indian Council of Forestry Research and Education [ICFRE] has identified 309 forest preservation plots of representative forest types for conservation of variable and representative areas of biodiversity. 187 of these plots are natural forests and 117 in plantations, covering a total area of 8,500 hectares.

A programme entitled ECO DEVELOPMENT of *in-situ* conservation of biological diversity involving local communities has been initiated in recent years. The concept of eco-development integrates the ecological and economic parameters for sustained conservation of ecosystems by involving the local communities with the maintenance of earmarked regions surrounding protected area. The economic needs of the local communities are taken care of under this programme through provision of alternative sources on income and a steady availability of forest and related produce.

World Heritage Sites

Programmes have also been launched for scientific management and wise use of fragile ecosystems. Specific programmes for management and conservation of wetlands, mangroves and coral reef systems are also being implemented. National and sub-National level communities oversee and give these programmes to ensure strong policy and strategic support.

Six internationally significant wetlands of India have been declared Ramsar sites under Ramsar convention. Additionally, eleven wetlands of national importance have been identified for intensive conservation and management.

Table 15.4. World Heritage Sites

Site	*Location*
Kaziranga National Park	Assam
Keoladeo Ghana National Park	Rajasthan
Manas Biosphere Reserve	Assam
Nanda Devi National Park	Uttar Pradesh
Sunderban National Park	West Bengal

The Tundra Range in Garo Hills of Meghalaya is a gene sanctuary for preserving the rich native diversity of wild Citrus and Musa species. Sanctuaries for Rhododendrons and orchids have been established in Sikkim.

Large mammal species targeted projects based on the perception of threat to them have been implemented.

The project Tiger launched in 1973 has succeeded in establishing and increasing the tiger population in the country.

Projects elephant, launched in 1991-92 aims at ensuring long-term survival of viable populations by restoring the lost and degraded habitats of elephants, mitigating man elephants conflicts and establishment of data base on the migration and population dynamics of elephants. It integrates the concern of improving the quality of life of people living around elephant habitats while maintaining viable population of elephants.

All these programmes, though focused on single species, have a wider impact as they conserve habitats and variety of other species in those habitats.

Several non-government organisations in the country are engaged in specific areas of target species-based *in-situ* conservation measures through National assistance or through a combination of assistance from both. An example of *in-situ* conservation for medicinal plants by Foundation for Revitalisation of Local Health-Tradition (FRLHT) in the southern states of the country needs mention here. In addition a large number of state government organisations are engaged in developing strategies and undertaking *in-situ* conservation. For example, Tropical Botanic Garden and Research Institute (TBGRI) in Kerala.

Gaps

The current Protected area network is unevenly distributed over states and biogeographic regions. Laddakh, South Deccan, Gangetic Plains, Assam Hill, and Nicobar Islands have less than 1% of their total area under such protection, though these constitute some of the biodiversity rich areas of the country. The Wildlife Institute of India based on a comprehensive review of the existing network argued for a need to identify new protected areas in different parts of the country.

Many National Parks and Sanctuaries are yet to complete legal procedures prescribed under the Wildlife (protection) Act 1972 and in several areas. Issues relating to settlement of rights are yet to be resolved. This impedes proper management of these areas, though steps are a foot to complete the process. Even the areas where legal measures have been completed require consolidation and strengthening of efforts with a system of sharply focused priorities. Formulation of management's action plans need to take these aspects into account. These projects are inadequate in essential equipments and professional staff and lack in maps and other requisites infrastructure base. Monitoring of activities and projects is not up to the expectation.

As of now, attention in the protected areas is increased on conservation of large mammals. Formally no attention is usually paid to conserve the plants in general and lower groups of plants and animals in particular. Consequently, there are gaps in information on several biological and managerial parameters.

Action Plans

1. Expand protected area network based on the recommendations of the Wildlife Institute of India.
2. Explore options for expanding/strengthening the network of protected areas through identification of buffer/ones corridors, etc.
3. Provide manpower, essential management tools, equipment, infrastructure facilities and requisite funds for ensuring effective management of protected area.
4. Expedite formulation of management plans for all protected areas based on Survey of National Research and socio-economic profiles in and around these areas.
5. Develop and implement eco-development projects adjacent to all the protected areas in order to meet the demands of local people and also to involve them in conservation efforts.
6. Establish self-sustaining monitoring system for overseeing the activities and effectiveness of protected area network.
7. Strengthen research programmes in protected area by involving Local Research Institutions, Colleges and Universities, so as to develop baseline data on biological and managerial parameters.
8. Device effective management and conversation techniques for the forest preservation plot to ensure

conservation of representative area of different forest types.

9. Ensure conservation of biodiversity outside the protected area network on private property, on common land and water bodies.
10. Minimize and eliminate activates leading to loss of biodiversity, including habitat destruction, over-exploitation, pollution and introduction of exotics.
11. Intensify measures for restoration of degraded habitats.
12. Strengthen and reinforce impact assessment requirements and parameters for location of activities in or around biodiversity rich areas.
13. Strengthen research programmes in designated areas for scientific management.
14. Reintroduction and establishment of viable populations of threatened species.
15. Intensify measures for restoration of degraded areas to meet demands of daily subsistence needs of local populations.
16. Start mechanisms for inter-sectoral consolations to identify programmes and activities leading to habitat destruction and over-exploitation of biological resources to design strategies for addressing these dangers.

Ex-Situ Conservation

Ex-situ conservation, which includes conservation of samples, genetic diversity (particularly representing endangered species) away from their field habitats.

Ex-situ conservation, using sample populations, is done through establishment of gene banks, which include genetic resource centres, zoos, botanical gardens, culture collections, etc.

Ex-situ conservation is the chief mode for preservation of genetic resources, which may include both cultivated and wild material. Generally seeds or in-vitro maintained plant cells, tissue and organs are preserved under appropriate conditions for long-term storage as gene banks. This requires considerable knowledge of the genetic structure of population sampling techniques, methods of regeneration and maintenance of varietal genepools, particularly in crosspollinated plants :

There are two ways or modes of *ex situ* conservation.

(1) Conventional method by seeds.

(2) Biotechnological aspect.

Conventional Method of Ex situ Conservation by Seeds

Germplasm conservation in the form of seed inmost convenient since seeds occupy a relatively small space. Their transport to various introduction centres and gene banks is also economical but the draw backs in conservation by seeds are:

1. Loss of viability over passage of time and susceptibility to insect or pathogen attack.
2. Inability to maintain distinct clones except for inbreed and apomict species.
3. Non-applicability to vegetatively propagated crop (Dioscorea, Ipomoea, Potato, etc.).

Note: For details on Ex-situ conservation see chapter: Biodiversity: Value Addition Through Biotechnology.

Other Methods To Conserve Degraded Areas

Reclamation and Conservation Through Afforestation

Afforestation of such areas and conservation of existing biodiversity will not only improve the site conditions, but will also improve the productivity and tree cover of the state. Prior to 1965, this area was devoid of any tree cover. Isolated growth of coarse grasses like *Sporoholus spp., Clitoris spp.* and a few weed plants like *Typha* etc., were observed in patches. Later on the entire selected area of about 38 ha was fenced with barbed wire, and arrangements were made to ensure effective protection against cattle grazing. One tube well was also made available for irrigation purposes during hot summer months.

Tree Planting

The trees, which are well adapted to xerophytic conditions, can be planted using following technique:

Prepare pits of 60 cm × 60 cm × 120 cm with a distance of 6 m × 6 m. The pits are filled with non-alkaline soil and 30 cm high mounds above the ground level is to be prepared. Farm Yard Manure (FYM) and gypsum are added as soil correctives in each pit at the rate of 3 to 5 Kg/pit. Irrigation is provided in summer months for two years.

The other methods are Silvipasture and natural regeneration of plant species in protected areas.

Conservation of Medicinal Plants

There is very interesting relationship between biodiversity and cultural diversity in the area of medicinal plants that is being lost equally because of the loss of cultural diversity associated with it also, of natural habitat and the various pressures that generally on biodiversity. Our folk traditions are ecosystem specific very diverse, very rich, and, there are the classical traditions systems like Ayurveda, Unani, Siddha, and Tibetan medicine. Therefore conservation of medicinal plants and culture is must.

Conservation efforts have been going on in different country is for almost three decades. The modern pharmaceuticals based on modern education systems looks at medicinal plants as resources for developing new drugs. That is the reason for an extremely limited effort for the conservation of medicinal plants.

The cultural traditions of thousands of communities are under serious threat. Undoubtedly, medicinal plants that have potential for making our rural communities self reliant in primary health care must be looked upon as a natural treasure. The medicinal plants that have been traditionally known to the people have not been explored potentially, everything is supposed to have medicinal value according to Charak, etc.

Total inventorisation of this has really not been done. In the folk tradition according to the Anthropological Survey of India, there are about 4635 different ethnic communities in this country, which comprise about 440 odd tribal communities. Tribal communities for the treatment of various diseases use about 7000 species of medicinal plants.

The Folk traditions are all ecosystem rooted. In fact a shloka in the classical Sanskrit text, which means that nature, is so endowed that every ecosystem has got the resources necessary for the purpose and survival of that ecosystem. So if one is living in Rajasthan one need not look for Himalayan flora in order to obtain medicinal plant resources. One striking example is that the plant *Ephedra vulgaris* which is found in trans-Himalaya. possesses bronchodilation properties and is only found in that ecosystem and of course commonly used by the local people as a herbal tea which is drunk several times during a day.

In fact, the ethnic utilisation of the biodiversity for medicine is absolutely tremendous. Everywhere, at every 100 Km, the local people have made independent appraisal of their local resources. Therefore, in terms of names of medicinal plants, the names vary at every 100 Km or so. In addition, in terms of use, in folk tradition there is tremendous diversity and here a number of plants species have been used by South Indian rural communities. For example, the *Centella axiatica* is a very well known plant both in folk tradition as well as classical traditions. Ethnic communities for medicinal or veterinary use have used the largest amount of local biodiversity. In a case study of the Western Ghats, the Madav Koh tribals live and of the way they are making use of their local biodiversity. 202 plant species are being used for medicine, 109 for veterinary, and in relation to other uses a high proportion of this use represents. It is a fact worth noting that in the diverse cultural folk traditions, herbalists for human and animal use have used the highest use of local biodiversity. The medicinal plants not only consist of herbs, but also of trees and shrubs.

Bonesetters also use large number of medicinal plants in their traditional practices. Therefore, the entire medicinal plant conservation strategy has been based so far on a knowledge system related to modern science. But if one gives recognition to these diverse cultural systems of use, then the value of medicinal plants to the rural communities is very substantial.

The alternation of habitats, the types of plants used in the classical tradition of medicine of Ayurveda, Unani, Siddha, Tibetan, etc., has been inscribed. In Ayurveda, there is an entire body of knowledge called Dhravya Guna Shastra, which is the indigenous knowledge of pharmacology; unlike modern pharmacology there is no category of chemistry in Dhravya Guna Shastra. Nevertheless, it is a very effective knowledge system for understanding the biological activities in plants. It is a holistic system of education. In terms of number about 14(K) plants are documented in various Ayurvedic texts. This is based fairly on vigorous study of these texts. There are approximately 342 plants in Unani, and about 328 plants in Ayurvedic that are largely used in the Tibetan system. In Homeopathy about 100 Indian plants are used while about 570 plant species are used worldwide. In the Allopathic system about 30 Indian plants are used in the pharmaceutical and aromatic industry.

There is large number of registered medical practitioners of traditional medicines than modern medicinal practitioners. There are a variety of reasons for certain weakness in the traditional system. But one weakness of which perhaps all are guilty is the general weakness throughout the intelligentsia, which simply does not understand the nature of the knowledge system that has been informing this large number of traditional practitioners.

Some more botanical aspects of conservation of medicinal plants are as follows. In the Ayurvedic medicine plants, the Tibetan family has got the largest number of species that have been traditionally used. A knowledge system under threat and the cultural diversity that is also under threat is Ayurveda. On one side it has the classical system of the medicine, which we call the Shastra, on the other side it is a modern science. There is difference in the foundations of the two Ayurveda, which is based on entirely different foundation principles and world view than modern science.

The largest threat to the biodiversity is from the lack of appreciation of the difference in the knowledge systems. The entire trend is to look at knowledge in terms of a mono-science as the only way to explain nature and, therefore, all the other ways in which the utilisation of the resources have been ignored. This is the most significant threat to biodiversity. To study the physio-chemistry and pharmacology of over 7000 plants, it may take roughly that many thousand years to do so. And so many thousands of herbal formulations have been tried and tested and validated by traditional knowledge systems. Over 15000 medicines are present, of herbal formulations for a range of ailments from common cold to complicated conditions. Lot of literature on the subject is also present.

A kind of narrow strategy has been adopted for the conservation of medicinal plants because of the lack of understanding and appreciation of the cultural diversity. For this reason, coordinated project through the Central Council of Industrial Research, which concentrated on the 45 species, was launched. Similarly, some agro-technology work has been done through the ICAR. Both of them concentrated on the 45 species that are important only for the modern aromatic industry with the help of a few non-government efforts. This all have been done for medicinal plants conservation till very recently.

These are some of the challenges for medicinal plants conservation. For appreciation the tremendous cultural diversity, it needs to establish a chain of *In situ* conservation sites all over India for protecting natural areas which have got a high amount of medicinal plant diversity. It needs to establish, in every single tehsil or taluka, ethno botanical gardens, which carry the plants that have been used by the traditional communities. More important than drug farms, it needs to establish a network of community nurseries, which will supply medicinal plant packages to the people to reinforce their traditions. And for this all the innovation, all the knowledge is available there in different ecosystems. For the sake of industry, there is a need to promote drug farms, instead of allowing the industry presently to go rampant and to collect material from the wild.

The Value of Medicinal Plants to India

India is being exhorted to earn revenues through export of medicinal plants. Billions of dollars are promised whereas herbal products can doubtedly be a source of revenue for India. It is often not realized that the economic value of medicinal plants to 100 million rural households in India is perhaps greater than what they can fetch in national and international markets. Hence, there has to be a balance in the approach and exports cannot be at the cost of domestic deprivations.

Half a billion people use neem as a toothbrush. The average toothbrush and a tube of toothpaste at a US drug store each cost more than a dollar. Does this make the toothbrush application a billion dollar business? If so, the neem value as a chew stick toothbrush is worth magnitudes more than the medicinal exports of India.

Guess-Estimated on Economic Value of Traditional Medicine

The majority of rural households use plants found in the local environment for managing their primary health needs. If one fixes a price on the use value in terms of the collection cost, the value of medicinal plants to rural households in India would work out to Rs. 100.00 × the number (100 million) of rural households in India = Rs. 10 billion.

If the exchange value of plants is calculated at a market price estimated at what the plant medicines in the market would cost an average family per year, the cost can be estimated at least at ten times the use value. This would make the economic value of medicinal plants used by rural communities in India to Rs. 100 billion.

It appears that the value of plants used by the Indian villagers is perhaps more than the value of medicinal plants utilized by industry in India as well as the value of plants that are exported.

Ecosystem Rooted

The ethnobotanical and other resources utilised in tradition are ecosystem specific and the local cultural diversity has a symbiotic relationship with the local biodiversity.

A striking example of this ecosystem linkage is seen in the case of the plant *Ephedra vulgaris,* which is only found in trans-Himalayan habitats.

This plant has broncho-dilation properties and local people traditionally drink a herbal tea made of the plant several times a day. The tea helps in easy breathing in the rarefied atmosphere of the Himalaya.

Some species of medicinal plants are on the verge of extinction.

Ethno biological Approach

Our mother planet, earth is endowed with an immensely rich variety of life forms which roughly consists of 3,00,000 Green plants, 8,00,000 Insects, 23,000 Fishes, 9,000 Birds, 6,500 Reptiles, 4,100 Mammals and few thousands microbes. These are nicely interwoven in the biosphere by natural laws forming ecosystem, where man has got supreme position and has exploited nature for his basic needs viz., food, clothing and shelter since his descent. In due course of evolution, he also unveiled the medicinal as well as magico-religious properties of plants and animals, through life long trial and error. Liking the oldest companions of man, plants and animals influenced culture and civilization significantly. Our ancestors imparted this knowledge to us through various folk tales and mythological stories. Lots of religious practices of worship and ceremonies associated with plants and animals are still observed. As far as biodiversity is concerned, it is the total variance of bio-entities present on the planet earth, which are altogether characterized by a single trail reproduction.

Therefore, ethnobiology and biodiversity represent the two faces of the same coin.

About 75% people of India reside in rural areas in close association with plants and animals. Besides these, there are over 400 different tribal and other ethnic groups in India, which constitute about 7.5% of India's population. Apart from these tribes, other forest dwellers, nomadic and rural people also have excellent knowledge about plants and animals. Thus, Indian region has a rich treasure of ethno biological heritage and there are more chances of biodiversity conservation through them. The people have little or no problem of food clothing and shelter while the health-care problems are extensively growing day by day. The inhabitants of remote areas, where health-care facilities are hardly available, get rid of their ailments by locally available plants and animals.

In developing countries like India, about 75% people are still dependent upon local bio-wealth to cure their diseases through crude drugs and strong magic religious beliefs. Therefore, ethnomedicinal and magic religious aspects of ethnobiology can better serve the purpose of biodiversity conservation. The only need is to teach the people about the real situation of loss of biodiversity, value of ethnobiology, their conservation and sustainable utilization for the present and future generations.

Ethnobiology Education Spectrum Approach

It includes various aspects, however, the important are:

1. **Awareness:** It intends to make the people conscious about biodiversity, its physical, social, aesthetic values and their link with ethnobiology.

2. **Practical life situations:** It aims to link the biodiversity with real-life problems through ethnobiology. The problems are location specific; therefore, they must be given priority.

3. **Sustainable Utilization:** The main object is to educate the people about the wise utilization and not the exploitation of biodiversity for the future well being.

4. **Conservation:** It is to educate the masses with the idea that all living resources have a limit of growth, therefore, for their fullest use, they should be conserved.

Implementation of Ethnobiological Education System (EES)

Formal Education System includes:

1. Primary Education
2. Lower Secondary Education
3. Higher Secondary Education
4. College Education and
5. University Education

In primary and lower secondary level education, the EES should be implemented in descending order; i.e., maximum emphasis should be given to sensitize the children about biodiversity and ethnobiology. The quantum of awareness should be maximum (75%), Practical life situation (20%) and conservation (5%). However, right from higher secondary onwards, the picture must be reversed and the quantum of conservation must be maximum followed by practical life situation and awareness. The contents to be used should be teaching, practical and action oriented field work.

Non-Formal Education System

This system can play a vital role in the conservation of biodiversity. This includes: (i) Adults education, (ii) Activities by Children viz. on the spot painting, modeling and poster designing contests, etc. (iii) Foundation courses to IAS, IPS, IPS and cadets of three wings of Armed forces need to be supplemented with EES. (iv) The Research Programmes associated with Flora and Fauna must also include their ethnobiological aspects.

Workshops and Camps

Regular workshops and camps for rural and tribal people should be organized at block and Nyaya-panchayat levels. The people can be educated and trained about EES by organizing lectures, displaying posters, slides, films, models and practical knowledge. Side by side, the propagules of plants supplemented with a booklet in local language with easy illustrations, including various lips associated with the regional ailments must also be distributed. This will not only solve the health problems but also add to the conservation of biodiversity by local inhabitants.

Papers and Electronic Media

(a) Media can play an important role in the implementation of EES and in turn conservation of biodiversity, because it is related to the common man.

(b) Celebration of Biodiversity Conservation Day or Week.

(c) All governments in states, union territories, academic institutions, schools, colleges, universities and voluntary organizations must organize suitable activities associated with EES, on a particular day or week of the year to attract the masses towards the occasion and its objective.

The problem of biodiversity conservation is not likely to be solved very easily in the near future until and unless a large proportion of human population comes to share the view. Also, no government or state can provide the complete healthcare facilities to all its citizens until they are self-dependent in this regard. However, ethnobiology, on the one hand, can conserve biodiversity through developing passion vis-a-vis species and provide complete healthcare to the society through ethnomedicine, on the other. The need is to explore the potentials present in plants and animals and to convey them to those who can really utilize and protect them for their own well-being as also future generations. Thus, a Maleria Medica of ethnomedicine and magico-religious plants and animals must be prepared region wise, which shall serve the dual purpose, of over coming the health problems and conservation of biodiversity. Through the combined efforts of various institutions, persons and government departments viz. education, forest and environment, agriculture, horticulture and health, the goal can be achieved.

Biodiversity-Means Of Production Or Product

For peasants and forest-dwellers, biodiversity has been the source of sustenance for basic needs such as foods, fiber, fodder, fuel, timber, shelter and medicine. The inbals and the farmers reproduce the necessary part of means of livelihood by planting crop each year.

New technologies by removing biological barrier transformed the means of production and product into mere material. The cycle of regeneration of biodiversity is thus replaced by a linear flow of free germplasm from farms and forests into corporate labs and research stations, and the flow of modified uniform products as priced commodities from corporations to farms and forests. Through technological innovations, biodiversity is transformed from a renewable into non-renewable resource. It does not produce itself; it needs the help of inputs. The manufacture of the products in corporate labs is regarded as production. The reproduction of the raw material by nature and Third World Farmers and forest dwellers is mere conservation. Biotechnology development thus leads to biodiversity erosion by way of converting the means of production or product into mere new material.

Politics Of Patents And Intellectual Property Rights

Biotechnological processes use life forms or derivatives to make or modify products or processes for specific use. Under Intellectual Property Rights (IPRs), transformed micro-organisms, plants and animals can be patented and become exclusive private property. The North has always used Third World Germplasm as a freely available resource and modified it. The issue of patent protection for modified life forms raises a number of unresolved political questions about the ownership and control of the genetic resource. Biotechnology does not create new genes, but merely relocates genes already existing in the organism. The advanced capitalist nations wish to retain free access to the developing world's storehouse of genetic diversity, while the South like to have the preprietory varieties of the North's industry declared a similarly public good. The North, however, resists this democracy. US has freely taken the biological diversity of Third World to earn millions of dollars of profits, none of which have been shared with Third World Countries, the original owners of the biological resource. For instance, an American Industry earned $8 million a year in 1962 simply by increasing the soluble solid contents of a wild tomato variety, *Lycopcrsicum chomreleivski* taken from Peru. None of these profits or benefits were shared with Peru, the original contributor of the genetic material. The Convention on Biological Diversity is also not clear on this score. Industrialized countries, particularly the US interpreted key clauses of the treaty in a manner that would protect the interest of its own biotechnology industries. This is a clear setback to the developing countries, who stand to lose the benefits due to them.

A developing country cannot match an industrial country in the absence of a proper biotech base, although the former may be far richer in biodiversity. However, the convention on biodiversity helped to place the subject matter of technology transfer and IPRs on the top of the agenda of policy and decision makers. Furthermore, access to genetic resources and transfer of technology are treated on the same plan.

On the issue of IPRs, the basic requirements of the Dunkel proposals is that inventions in all branches of technology shall be patentable, whether products or processes, if they meet the three tests of being new, involving innovative steps and being capable of industrial applications. It has also been provided that micro-organisms will be patentable. In respect of plant varieties, there is a separate obligation to provide them protection by patents or by an effective surgeries system, or by a combination of the two. So far we are concerned, seeds are also not patentable in India today, and we do not have any intension of changing this system. However, we will adopt our own system for the protection of plant varieties under which we may provide certificates for plant breeders right. The farmers rights include the using of the seeds for their own needs or for exchange in the village community according to their traditional custom. Since farmers right will be fully safeguard under system of protecting the plant breeders right, there is no truth in the declamation that the farmers will not be able to reserve the seeds for their own use and that they will have to buy seeds every year from multinational companies. Furthermore, India is not in favour of the patenting naturally occurring life forms germplasm.

The extension of IPRs to plant varieties either in the form of patents or in the form of Plant Breeder's Rights is bound to result in increased in prices of seeds, greater domination of agriculture by multinational companies and slower diffusion of new varieties. These would be in sharp contrast to the experience of the Green Revolution where the new varieties of seeds evolved by the government institutions percolated down to the fields in a short span of time with very little cost to the individual farmer. India calls for the removal of distortions in the IPRs regimes in areas related to prior existence of knowledge in nutrition.

The South India had not merely preserved much of its precious biogenetic resources, but also the knowledge and practices about their optimum and sustainable utilisation. Access to these resources have to be regulated and carefully exercised, in keeping with the objectives of the convention and with due compensation to such people who have preserved their resources. The decision of the CSD to include in their medium-term programme, the knowledge, innovations and practices of indigenous and local communities is an important step in the direction of the protection of traditional knowledge and practices of the indigenous and local communities relevant to conservation and sustainable use of biological diversity.

Bio-safety Protocol

India has demanded a clear comprehensive and legally binding international protocol on bio-safety under the convention on biodiversity. Addressing the first meeting of parties to the Convention on Biodiversity, at Nassau, Bahamas, on December 8, 1994. Mr. Kamal Nath demanded immediate and adequate safeguards against hasty experimentation and use of genetically modified organisms, since these have

unimaginable act of driving behind. He feared that the developed world could become a playground for experimentation, of such genetically modified organisms, and it could only be checked through a legally binding agreement.

It is the direct use value of biodiversity, which is highly valued by local people in developing countries because of the subsistence nature of economy. People living in abject poverty, destitution, and miseries are neither bothered about sustainability nor option, existence or other values of biodiversity. International organization and the developed world on the other hand are concerned about the option and existence values of the biological resources since their approach has shifted from Quantitative Growth to Initiative Development.

Strategies For The Future

Development of both the individual and the social, environment are two sides of the same coin. Due recognition is needed for global environmental risks while designing any environmental strategy. Environmentally and socially disruptive economic behaviour must be changed if we intend to provide decent livelihood to all citizens of the present and future generations and this change can only be brought about by discarding environmentally insensitive forms of development. Rather, the stress should be on qualitative development based on efficient use of energy and natural resources, recycling of end products and minimizing the use of non-biodegradable products.

In a country with the increasing human and livestock population, concrete actions and sound strategies are easy to formulate but difficult to implement. Human beings have a tendency to prefer present goods to future goods and in the process, overlook the side effects. To sustain 1.5 billion people in India by the year 2025 heavy financial investment and state-of-the-art technology will be needed in agriculture, water resources, infrastructural development and so forth. Ecological balance will definitely be adversely affected to create liquidity despite our best efforts not to do so. Primary and secondary forest industries will use more wood, cork, bamboo, cane, gum, resins, essential oils, seeds, medicines, etc., for home consumption as well as export in a liberalized economy.

Despite all adversities one has to adopt viable strategies for the conservation of species as this will only guarantee our future and lead us towards the goal of sustainable society, sustainable economy, sustainable use and sustainable development. The strategy should be aimed at policy makers, all Government and non-government organization, private organisations, individuals and intergovernmental organisations. This will help in achieving the objectives of biodiversity conservation and sustainable development. Any conservation measure has to be carefully planned so that:

Environmentally and socially disruptive behaviour is altered to provide a decent living for all.

1. The production pattern is socially equitable.
2. A human centered, knowledge intensive civilization is built.
3. A balance is maintained between all forms of capital i.e., human, natural, physical and financial.

Some of the measures suggested as a part of long-term planning are as follows:

Carrying Capacity of The System

Carrying capacity varies from region to region depending upon population pressure, their dependence on food, water, energy, raw material requirements, waste production and import and export. In order to live within the carrying capacity the society will have to change the life style and consumption pattern in such a manner that causes least damage to the ecosystem. Over population in a particular region has to be avoided and excessive pressure on land must be reduced. It calls for changes in the present land-use policy, overhauling of land tenurial system and population control. Forest land will have to be used in such a way that maintains their biodiversity, productivity, regeneration capacity, vitality and their potential to fulfill, now and in future, relevant ecological, economic and social functions.

Enhancement of the Protected Areas

One of the best means of protecting the natural forest is through the establishment of protected areas, an effort that has taken more than a century to evolve from the concept of reserve forest to national parks and biosphere reserves. These protected areas serve three basic functions viz., benefit, education and recreation of the people.

Although, India has made steady progress in the management and expansion of the protected areas, yet, the problems are being compounded with time. Mere establishment of protected area does not ensure socio economic development and species conservation. The management has to take timely action for the speedy disposal of the settlement problems. Local people who consider themselves as the primary stakeholders must be involved in the use and management of the resources. Socio-economic well-being of the people should be one of the top priorities for minimizing the resource dependency on the core of these protected areas. Biological resources in the protected areas can be assigned consumptive and non-consumptive values so that the conservation of biological diversity can be linked to sustainable socio-economic development.

At the same time, there is a need for change in the attitude of the field staff towards the interests of the local people. This conflictual relationship has to be transformed into cooperation and partnership. Field staff will have to appreciate that with inadequate workforce, prohibitive management strategy is impossible to implement. Therefore, in the long run a conciliatory interactive and motivational approach seems to be the best.

Enhancement of protected areas will need effective management with clear objectives for each area. This will require increased investment in human resources, education and research.

Stress on Integrated Development

It has been realized that a direct stress on conservation measures and neglecting the economic development of the people in and around the protected areas leads to impoverishment, and initiates the vicious cycle of subsistence, emigration, land fragmentation, degradation and deforestation. According to the new concept, people will have to change their consumption pattern and find alternatives on their own, once their purchasing power increases. The only fear in this integrated effort is that short term conflicts may arise in the absence of appropriate policy measures causing hardships to the people. It is therefore essential that conservation, rural development and compensatory incentive programmes should be organised simultaneously.

Research Impetus

There are five main areas, which need urgent attention:

Forest management

Forest management has undergone rapid changes in the decades of eighties and the nineties. Geographical information system and remote sensing techniques have the potential to provide a large amount of information on forest resources, growth and biodiversity.

Restoration of degraded lands

Research efforts should concentrate on enhancing the soil micro fauna and fertility status so that sustainable production can be achieved by adopting agroforestry systems that can support the economic development of the region. Efforts should be made for genetic improvements and dissemination of multipurpose tree species that can grow in degraded areas.

Research in harvesting, processing and marketing

There is a need to develop low impact systems, drying and conversion processes for non-commercial abundant species, especially the NTEP's. Marketing management has been one of the weakest sectors of forestry and needs new directions and support.

Policy institutional and socio economic research

Socio-economic objectives are best pursued if the policies regarding land-use, rehabilitation of degraded sharing, land and tree tenure participatory planning and resource management are appropriate to the sustainable management and biodiversity conservation.

Research in Biotechnology

Sustainable development and biodiversity conservation in the 21st century will depend upon proper exploitation of economic potential of genetic diversity using biotechnology for the production of goods and services required by the increasing population. Foresters will have a dual role in future: one as *in situ* conservationists and the secondly in *ex situ* maintenance of biodiversity for economic use (also see chapter on: Biodiversity: Value Addition Through Biotechnology in this book).

In addition, social and anthropological research will also be required to determine individual and community needs and perspectives of forests, trees and environment. This will have immense impact on the participatory planning and management processes. Last but not least, the concept of conservation and development should permit our thinking and perception.

Protected Areas for Biodiversity Conservation and legal Position

The depletion of biodiversity is an alarming problem in all over the world. The World Conservation strategy suggests that the initial efforts of biodiversity conservation should aim at establishment and maintenance of a network of Protected Area (PA) system by making policy changes involving local people in the PA management and mobilizing financial resources for their conservation and protection. The strategy also lays down that considering the stress on environment, all countries should aim at earmarking 10% of their land area for PA network for biodiversity conservation.

For quite sometime, especially after the 1991 amendments of Wildlife Protection Act, some of the States have been quite vocal about the rigid protection that is being attempted to be given to the protected areas denying the local people's access to biological resources of PAs for their use and sustenance. While in the first instance, such demands or hue and cry may appear realistic, a deep look into the issues and objectives will reveal that there is no anomaly in the PA management concept that is being attempted in India today.

We have to keep in mind that protected areas make vital contributions to the conservation of the world's natural and cultural resources. Values range from retention of representative samples of natural region, its wild denizens and the preservation of biological diversity to the management of environmental stability of the surrounding region viz., the fertility of marginal lands, preservation of a wide variety of genetic resources, maintenance of quality and availability of water, climatic amelioration, etc. They also provide scenic and aesthetic surroundings and offer facilities for recreation and tourism.

A well-managed forest may be able to provide sustainable supply of a few forest tree or plant species, or a group of species in perpetuity. But there is no model of management yet known which can be applicable within the short span of human life that can replicate the biodiversity of a given area in totality. It is because of this that the conserved PAs of India attract attention of global community and there is a clamour for visits to these areas for the purpose of education and enjoyment of the people. Thus, the range of services and the values that the protected areas offer cannot be compared with other forest areas and therefore, the management objective of protected areas cannot be the same as applicable to other areas in general.

Unfortunately, the PAs all over the developing countries are now subjected to tremendous pressures, especially because of the burgeoning human population followed by use, over use and sometimes abuse of such areas, leading to destruction of the biological resources in the name of sustenance of the local people. In these countries the glare of the Western World and the profligate life style practiced in the developed countries have led to more and more resource use leading to consumerism of higher order. This is unsustainable, given the high population density that exists in such developing countries, especially in the continent of Asia including India. Demands are coming from various sectors that the PAS should also be opened to human use if their long-term sustenance is to be guaranteed. A question, therefore, arises: does our past experience indicate that when the people enjoyed full right or ever ownership over the forest resources, could they or did they protect the forest resources under them, particularly under Indian conditions, in a better way than the Forest department? If the answer to this question is negative, then it is an urgent need of the hour to live PAs to preserve the rich biological heritage.

The Commission of National Parks and Protected Areas (CNPPA) of IUCN has been deliberating on this issue since 1978. In the Steering Committee meeting of the CNPPA held in 1993, a consensus was reached to have the following set of categories for the PAs. These are given below with the broad management objectives.

Category I *Strict nature Reserve/Wilderness Areas:* Protected areas managed mainly for science or wilderness protection.

Category II *National Park:* Protected areas managed mainly for ecosystem conservation and recreation.

Category III *Natural Monument:* Protected areas managed mainly for conservation of specific natural features.

Category IV *Habitat/Species Management Areas:* Protected areas mainly for conservation through management intervention.

Category V *Protected Landscape/Seascape:* Protected areas managed mainly for landscape/seascape conservation and recreation.

Category VI *Managed Resource Protected Area:* Protected areas managed mainly for the sustainable use of natural ecosystems.

In India, PAs belonging to (categories II, IV, V and VI) are being managed. Category I corresponds to the sanctum, sanctorum of the PAs and Category III sometimes occur incidentally in various PAs, but very often these areas may be occurring outside the conventional PAs also. National Parks in India fall in Category II, Sanctuaries come under Category IV and Biosphere Reserves come under Category VI.

It can be seen that in National Parks, the main objective should be ecosystem conservation and recreation. The sanctuaries should fall under habitat/species management areas, which are to be conserved through managemental intervention. Category V are areas basically maintained for the protection of catchment areas of rivers, sea coasts, and for recreation.

Thus, joint management of protected areas has to be in the form of ecodevelopment around PAs, Through resource sharing to elicit people's participation. This has to be incidental and ancillary to the management system and not the objective of the system itself.

The National Parks should be kept inviolate so as to maintain biodiversity and continue ecological processes. In sanctuaries, however, some amount of activities is inevitable and there are certain provisions in Wildlife Protection Act to this effect.

Section 29 of the Act says that no persons shall destroy, exploit or remove any wildlife or its habitat within a sanctuary except under, and in accordance with the permits granted by the Chief Wildlife Warden, and no such permit shall be granted unless the State Government is satisfied that such destruction/ exploitation or removal is necessary for the improvement and better management of wildlife therein. A number of our protected areas are having commercial and pure plantations of Teak, Eucalyptus. Ailanthus, Poplar, Anthocephalus, etc., under which, there is hardly any understorey or desirable undergrowth. These are certainly not good as wildlife habitats as they neither provide adequate cover, nor fodder which are essential requirements of wild animals. It would be beneficial to carry out heavy thinning of such plantations and raise desirable species, which could provide good fodder and cover to wild animals. This can be feasible in the sanctuaries. The local people living in the fringe areas of such sanctuaries can be employed gainfully in such activities and atleast a part of the produce coming out of such thinning can be distributed to them at concessional rates, or under some mutually agreed terms and conditions. Unfortunately, this is not being visualized in most areas. Similarly there is no bar to take up some activities in open and degraded portion of PAs to develop it as ideal habitat for wild animals.

Very often arguments are put forward that in protected areas no rights of local people can be allowed. This is also not correct as Section 24 (2) (c) of the Wildlife Protection Act has a provision to allow, by the collector in consultation with the CWLW, the continuance of any right of any person in or over any land within the limits of the sanctuary. However, obviously the objective is to allow only such rights, which are compatible with the long-term objectives of conservation of the protected area. The actual situation is that in most of the cases the rights were extinguished at the time of declaration of the reserved forest but were not enforced because the objective of ecosystem conservation or the conservation of biological diversity was not a high priority in the management of our forests earlier. In other places, even if the sanctuaries and national parks were declared, the process of section 19-25 of the WLPA was not followed which extinguishes or admits the rights of local people. It is only when the protected areas were transferred to wildlife wing and serious attempts were made to control the illegal collections or the processes under section 19-25 were initiated, the hue and cry blaming that the wildlife laws are creating hardships was raised.

However, it has to be agreed that demarcating out of the protected area system does to some extent, deny general access of the people into such areas and unrestricted use of forest produce in the same pattern as is allowed in other categories of forests.

It is often said that the traditional life style of the trans-humans was earlier not only ecologically compatible, but also helped in harmonious development of both the indigenous people and the ecosystem. Without entering into the argument whether this was so in all cases earlier, it has to be remembered that in the earlier era forests were plentiful and the population was less. Over centuries, the forest cover has declined and the human population has increased. Thus, today there is a big gap between the demand and the supply capability of the forest/wildlife. Therefore, if the biomass production from the forest areas, or outside is not increased or the alternatives to existing biomass uses is not found out, it will be impossible to meet the demands of the local communities. Apart from that, access to modern living and the profligate style of life has also greatly affected the living style of the people living in or around forest areas which has resulted into increased consumerism and increased pressure on the existing biomass much more than that in past.

It is also advocated that the local communities are gradually getting impatient and alienated to the protection measures. While it is true in some cases, this is not the picture everywhere. In most of the cases, adequate attempts have not been made to educate the local communities about the services that forests and wildlife render towards their sustenance in the long run, and to look into their genuine problems. In absence of such extension measures, some unscrupulous people and opportune seekers take advantage of the ignorance of the illiterate local people and instigate them to resent against any control measure which apparently appears to cut down their access to traditional use of forest and wildlife. On the other hand, where local people have been convinced of the utility of forest and wildlife towards their long-term sustenance, the same community which was damaging wildlife and its habitat earlier are now imposing restrictions on the use of habitat and the habitants themselves. Thus, the common people need to be educated about our culture and tradition, and the differences in the scenario that exist today and that of the past, and the reason why the erstwhile traditional living style, which had not adversely affected the forests till about fifty years back, cannot continue in the present circumstances.

If this is done, a lot of misunderstanding and resentment which is there in a number of areas can be resolved and harmonious growth of people and protected areas can be ensured without bartering the forest produce beyond the limit of its sustenance and management objectives.

We have to remember that the forests within reserved and protected forests, as well as protected areas, exist today because of the legal protection offered to such areas. The need of the hour is that the provision of law should be strengthened by change of people's perception towards conservation, combined with certain measures, which will also ensure some amount of benefit to the communities and individuals. Some amount of hardships to local people may be caused initially, but these can be partially mitigated through launching of eco-development programmes and fulfilling the biomass needs of the local population. This can be achieved by creating the required forest resources in lands outside the protected areas, and by the introduction of appliances and techniques for efficient utilisation of energy and biological resources to reduce wastage.

The eco-development of each PA should have a local committee, which will decide as to what would be the package of such eco-development programmes. The sense of cooperation within local population would automatically develop for participating with the PA manager not only in the eco-development plan, but also for the entire management planning of the PA.

Joint management in the context of protected areas should essentially cover participation in the protection of all zones and responsible use of resources from the buffer zone of the PAs and multiple-use zones around them. In fact, amount to improvement in PA management because once the degrading factors are brought under control; the restoration of productivity will be brought about by nature itself.

Biodiversity Conservation Strategies For The 21st Century

Diversity of species in natural habitats is high in warm areas and decreases with increasing latitude and altitude.

Diversity is usually higher in high rainfall zones than in drier areas. That is precisely the reason why tropical moist forest contain half of the world's species though they occupy only seven percent of the land area.

The word richness indicates the number of species present in a designated area, Whereas 'evenness' stands for the relative abundance of each species, It means that a pristine forest and a seriously degraded site may have identical number of species but the evenness will decide the degree of diversity.

Although species extinction is a natural process that will continue without human intervention. However, it is a fact that present loss of species is one thousand to ten thousand times faster than the erstwhile natural rate of extinction. Developing countries of Europe and North America have depleted their forest resources, which as it is, had limited biodiversity. Most of these nations have undertaken large-scale commercial plantations on forestlands and shall have to rely heavily on the tropical nations for their future requirements of plants and animal species. Developing countries in the tropical zones are rich in biodiversity, but face the dilemma of its conservation vis-a-vis economic development.

According to a rough estimate there are 5-30 million living species on this earth of which, only 1.4 million have been identified so far (William, 1988). Each species has nearly 4,00,000 genes and virtually no two members of the same species are genetically identical. Most of the quoted global extinction rates are derived from predicted rates of habitat loss and estimates of species richness and evenness in different habitats. Although, in the present global scenario the loss of crop varieties and livestock breeds is of little significance but genetic erosion of these populations is a cause of serious concern since the wild varieties may provide valuable traits like disease resistance, nutritional value, hardiness, etc.

Pic. 15.1. A migrants' colony at Kattebains near Vasco de Gama.

16

BIODIVERSITY CONSERVATION: INDIA SCENARIO

Gujarat, located in the western region of India, has a variety of climatic and edaphic conditions. Inspite of continuous onslaught on the biological resources, the overall range of biodiversity of Gujarat is still unmatched. Gujarat has many kinds of natural ecosystems due to varied environmental, climatic and edaphic conditions. Some important ecosystems in this state are deserts, grasslands, scrublands, agri-fields, wetlands, forests, marine ecosystem etc. This state has got a large population of tribals dependent on the natural sociological resources for their livelihood. The southern and eastern Gujarat are primarily tribal areas whereas in Saurashtra and Kutch, there is a large population of nomadic pastoralist community called Maldharis, whose migratory lifestyle effect the floral and faunal diversity of the state.

Gujarat is situated in the Central Western part of India, with an area of 1,96,020 Sq. Km and about 1,600 km long coast. The eastern hilly region is formed by southern, western and northern extension of Arawalli, Satpura and Sahyadri (Western Ghat) Mountain ranges, respectively, and vast extents of saline deserts form a part of this state. The broad spectrum of the eco-climate is defined by rainfall varying from 250 mm in North-West to 2,000 mm in southeast. Seasonal variations of tropical monsoon climate are distinct. On the coast, there are two important sheltered gulfs viz. Gulf of Kutch and Gulf of Khambhat.

Of the ten bio-geographic zones of India, Gujarat has parts of four zones, namely, zone 3, 4, 5 and 10. The provinces of this zones in Gujarat are 3A-Kutch, 4B-Gujarat-Rajwara, 5A-Malabar coast, 5B Western Ghat Mountains and 10A-the West-Coast. Based on the physiography, rainfall pattern and the presence of some indicator species of Western Ghats (considering the results of faunal and floral studies carried out by M.S. University team in the Narmada valley region in Gujarat), the Province 5A and 5B have been extended and shown as reaching up to the left bank of the Narmada river.

Gujarat Ecology Commission initiated a research programme of preparing a Status Report of Biodiversity of Gujarat (based on scientifically recorded species and their distribution) with the following objectives:

1. To discover and describe species diversity (i.e., describing biodiversity).
2. To analyze and synthesize the information derived from the Research Programme into a predictive classification system that reflects the history of species (i.e., understanding biodiversity).

This preliminary work of compiling secondary information on biodiversity of Gujarat would provide a basis for future studies and plans. India has over 1,26,188 species of bacteria, fungi, plants and animals already described, of which, nearly 7,850 species are recorded from Gujarat. These species occur on land and in freshwater and marine habitats or as symbionts in mutualistic or parasitic state with other organisms.

Table 16.0. Number of Recorded Biota in Gujarat (based on available data)

	Taxon	*No. of Species in Gujarat*
FLORA	Bacteria	-
	Algae	1933
	Fungi	164
	Lichens	-
	Bryophyta	
	Pteridophyta	16
	Gymnosperms	1
	Angiosperms	2198
	TOTAL	**5040**
FAUNA	Protozoans	255
	Porifera	69
	Cnidaria	78
	Ctenophora	-
	Platyhelminthes	
	Nematoda	-
	Rotifera	87
	Kinoryncha	-
	Gastrotricha	
	Taxon	**No. of Species in Gujarat**
	Acanthocephala	-
	Sipuncula	15
	Mollusca	350
	Echiura	11
	Annelida	69
	Onychophora	-
	Arthropoda	743

Phoronida	1
Bryozoa	42
Entoprocta	-
Brachiopoda	1
Chaetognatha	-
Echinodermata	15
Hemichordata	
Protochordata	-
Fishes	364
Amphibians	19
Reptiles	78
Birds	454
Mammals	69
TOTAL	**7843**

There are about 1,933 species of Algae reported from Gujarat, out of which, 210 species are marine. Marine algae are known since long as a source of food, fodder and fertilizer. Barring a few, all species and all genera found on the Gujarat Coast are cosmopolitan. Out of the 23,000 species of fungi reported from the Indian region, only 165 species are documented from Gujarat. Similarly, very scanty information is available on the groups like Bryophytes, Pteridophytes and Gymnosperms.

The angiospermic flora of Gujarat is mostly varied in existence and composition. There are 2,198 species of higher plants belonging to 902 genera and 155 families which represent 12.91% of the flora of the country. So far as the floristic explorations are concerned, several workers have intensively studied the South Gujarat region. By and large, the central part of the State, part of North Gujarat, West Saurashtra and Eastern Kutch remain virtually unexplored. 62 species of Angiosperm, and Gymnosperm plants have been given different rarity stains. Majority of them are distributed in the semi-arid regions of Kutch.

Twenty-seven species of mangroves are listed from Gujarat. Earlier reports show that there was a luxuriant growth of mangroves at Sikka, Salaya, Jodiya, Jakhau, Modhwa and elsewhere on the Gujarat coast. However, drastic reduction has been noticed in the recent years due to extensive cutting for timber, fuel and fodder.

The ethnobotanical studies carried out by several investigators have identified nearly 1,200 plants species used by tribals for meeting their multifarious requirements.

The present survey of fauna has revealed that though a good deal of work has been carried out in floristic explorations, no comprehensive efforts have been made to explore the faunal wealth of Gujarat, barring the vain and mammalian species. Altogether, references of 3,462 species of animals belonging to different tax recorded from Gujarat have been collected. The fauna recorded from Gujarat is 4.54% of the Indian faunal population.

Euglcnoids and Amoeba are the protozoan groups studied systematically in Gujarat while the remaining phyla namely, Ciliophora, Apicomplexa, Microspora and Myxo remain untouched. Our knowledge on the Calcareous and Hexactincilid sponges of the Indian waters is still confined to the classical monographs that appeared in the pre-independence period. According to the Zoological Survey of India (1991), the sponge fauna of the West-coast of India is rather poorly worked out compared to that of the East Coast. The freshwater sponges recorded from Gujarat are less than ten species.

Among the colorful marine Cnidarians, only the corals are studied systematically along the Gulf of Kutch while other two classes of Cnidarians (Hydrozoa and Scyphozoa) have yet to gain attention. All the 37 species of corals in the Marine National Park are reported to be in the endangered/threatened category.

Platyhelminthes constitute one of the major phyla comprising about 17,500 known species out of which, 1,622 species have been recorded from India. However, no attempt has been made in Gujarat to study the taxonomy of this phylum. One species of Planaria is included in the vulnerable category by the Marine National Park authorities. Records indicate that among Aschelminthes, only Rotifers have been systematically studied during the sixties in Gujarat.

The Molluscan fauna of Gulf of Kutch is fairly well known. About 350 species have been identified from this coast. However, the information available is mainly focussed on the edible marine molluscs as it is one of the important marine foods in Gujarat. According to the Marine National Park authorities, many Molluscan species are threatened due to over exploitation in the Gulf of Kutch. In contrast to marine Molluscs, little information is available on the freshwater Molluscs in Gujarat.

Annelids constitute one of the largest phylum with 1.093 species reported from India of which, only 69 species are reported from Gujarat. The studies were mainly on marine Polychaete worms along the Gulf of Kutch. Little is known about Oligocheates and Hirudenians. The need of systematic studies of these two classes is obvious.

The other largest phylum, Arthropoda, is represented by more than 50,000 species in India. However, only 743 species are reported from Gujarat. Some isolated studies on scorpions, pseudoscorpions, crop pests and mites have been carried out. Most of the forest areas in Gujarat are unexplored from the entomological point of view. The recent studies conducted by a team of M.S. University, Barodu in Shoolpaneshwar Sanctuary suggest that many more species can be reported from the forest areas of Gujarat if thorough study is carried out in these regions.

The Bryozoans of Gujarat (42 spp.) comprise a good proportion of the total 170 species recorded from India. On the other hand, of the total 765 species of Echinoderms reported from India, only 15 species are known in Gujarat.

Hemichordata is a group which is not studied adequately from Indian waters, and there is an urgent need for taxonomic studies, particularly on the families Protobalanidae, Cephalodiscidae and Pianctospaeridae.

In comparison to the invertebrates, the vertebrate fauna has been studied fairly well in Gujarat. The fish fauna is represented by 364 species, of which, 19 species are in the endangered/threatened category. So far, the non-edible fishes and freshwater fishes have not received adequate attention in taxonomic surveys. Only 19 known species, which are mostly confined to the southern districts of the state, represent amphibian fauna of Gujarat. Appropriate monitoring and conservation strategies need to be developed to preserve these species. 78 species of reptiles are reported from Gujarat and 8 of them are placed in the Red data Book of ZSI. Many reptilian species are facing severe threat due to habitat destruction.

The avian fauna of Gujarat is unique and is represented by 454 resident/migratory birds. 8 species are placed under the endangered and 5 species in the vulnerable category in Red Data Book of ZSI. In addition to this, 11 species of birds are listed as endangered/threatened by the Marine National Park authorities. The Mammalian fauna of Gujarat comprises of 69 species, which is 18.49 per cent of the total Indian Mammals. Four species are featured in the Red data Book as endangered while ten species are in the vulnerable category.

Analysis of the livestock population of Gujarat has shown that over a period of ten years (1982-1992) the livestock population growth rate is 7.16 per cent. District-wise population growth of cattle, buffaloes, sheep and goat has also been worked out. The data indicates that measures are required to study the carrying capacity of the districts as well as the feeding practices in some districts. Altogether, the number of indigenous breeds of Gujarat is far higher than those of the exotic verities.

Based on the available data, an attempt has also been made to categorize the existing forest areas into unique and rich areas of biodiversity.

The Gir forest area, Shoolpaneshwar Wildlife Sanctuary, Marine National Park and Dang forests are classified as Unique and Rich biodiversity areas. Vansda National Park, Vajpur forests of Surat district, Ratanmahal, Vijayanagar forests of Sabarkantha district, Velavadar National Park, Girnar forests, Grasslands and Rakhals of North-Western Kutch are categorized as biodiversity rich areas. Among Unique biodiversity areas, The Flamingo City in the Great Rann of Kutch, North Arid Zones of Kutch supporting the Spiny-tailed lizard (Uromastix hardwickii). Wild Ass sanctuary in the Little Rann of Kutch and Gaga. Great Indian Bustard sanctuary are included due to the presence of unique species of animals.

Management Suggestions

It is essential for a Wildlife Manager to study biodiversity and its habitat before attempting any addition/alterations. The following suggestions are made for conserving the biodiversity of Gujarat:

Captive breeding and reintroduction programme should be taken up for rare and endangered animals and birds of Gujarat. Sakkarbaug Zoo of Junagarh is already doing commendable job for breeding Asiatic lions, Indian wild ass and other important animals and birds. Reintroduction programme should be taken for such important species, after carrying out proper study. Government of India with the help of Wildlife Institute of India, Dehradun have taken up introduction of Asiatic lions in a second home at Kuno Panpur Sanctuary in Madhya Pradesh.

Conservation and protection of existing natural forests of Gujarat, particularly that of South Gujarat, Panchmahals, North Gujarat, Gir and Girnar.

Environmental awareness and nature education programmes have been a great success in Gujarat. It should be extended to other areas. Nature conservation societies and clubs can be of great help in conserving biodiversity, as these bodies are an important link between the Government and people around forests, mangrove, grassland, wetlands and other Protected Areas.

Extension of lion habitat can be attempted on the lines of GGE concept by taking up suitable habitat management and proper and sufficient water and development of prey base.

Intensive research should be carried out in key biodiversity areas to study the habitat, ecology and food requirements of wild animals, biotic pressures etc. Marine national park area in Gulf of Kutch needs more immediate attention.

Key biodiversity areas should be protected from cattle grazing. In such areas, ecodevelopment programme (as being taken up in Gir) would be of immense help. In semi-arid regions, integrated watershed development works should be taken up on large scale.

Natural bamboo breaks/bamboo forests of Rajpipla form excellent habitat for several wild animals and therefore, need to be conserved.

Preservation Plots should be protected and conserved with vigour and continuous study should be carried out to monitor the changes in the biodiversity status of these areas.

Preserving large sized trees for flying squirrels, grey hornbill, birds of prey and Owls.

1. Protection and improvement of vast grassland ecosystems of the state especially non-reserve Vidis.
2. Shifting of some of the local Agariya salt works in small Rann of Kutch, which block the local migration of wild ass.

Study and conservation of sloth bear habitat in hilly-forested patches of Jessore, Ratanmahal, and Rajpipla forests. For the enlargement of Sloth bear habitat, Girnar and Jambughoda can be taken up on experimental basis as these areas are suitable to sloth bear population.

PROTECTED AREAS IN GUJARAT

1. Gir Lion National Park and Sanctuary

Location: Junagarh and Amereli district

Area: Sanctuary 1153.42 sq.kms.

National park 258.71 sq.kms.

Total: 1412.13 sq.kms.

The last abode of Asiatic Lions (Pantheru leo persica) in the entire world. The habitat represents the best example of the dry deciduous forests in the country. It has got the highest concentration of the two largest predators in the country i.e. Asiatic lions (304) and panthers (230). Other wild animals include cheetal, sambar, chowsingha, chinkara, wildboar, nilgai, ratel, rusty-spotted cat, pangolin, crocodile, etc. It is one of the best managed National Park and Sanctuary in the country.

2. Marine National Park and Sanctuary

Location: Sea coast in Jamnagar District

Area: Sanctuary 458 sq.kms.

National Park 163 sq.kms.

National Park area is included in Marine Sanctuary area

It is the first Marine National Park and Sanctuary in the country with rich marine floral and faunal diversity. There are more than 40 species of Corals and 70 species of Sponges. Bonnelia, which is recorded only at three placed in the world, is found here. Three of the world's six marine turtles (leather back, green and Oliver Ridley's turtle) are found here. Besides, large marine mammals like whales, dolphins and dugongs are also seen in the Gulf of Kutch. Other marine life includes sea horse, sea anemone, star fish, crabs, coral fishes, sea hare, octopus, squid, sea shells etc. According to late Dr. Salim Ali, the country's best aquatic and marine bird concentration and one of the best mangrove forests are found here.

3. Velavadar Blackbuck National Park

Location: Bhavnagar district

Area: 34.08 sq.kms.

A grassland of very high ecological value in the saline Bhal region has almost flat terrain. It falls in the low rainfall region of the semiarid Saurashtra. It has the largest concentration of blackbucks along with the only predators of this area, i.e., wolf. With effective improvement of grassland status in the park, the area now has perhaps the largest concentration of lesser floricans (above 42) during breeding season.

4. Vansda National Park

Location: Valsad district

Area: 23.99 sq.kms.

This contains moist deciduous forest which falls in the western Ghats and contains wildlife such as panther, cheetal, sambar, wildboar, barking deer, racket tailed drongo and orioles etc. Although, the habitat is well suited for wild animals, the animal population is very low.

5. Nal Sarovar Bird Sanctuary

Location: Surendranagar District

Area: 120.82 sq.kms.

The largest natural wetland habitat in the state which attracts large number of winter migrants such as ducks, geese, storks, spoonbills, flamingoes, plovers, coots, cormorants and pelicans etc. A heaven for ornithologists, naturalists and bird watchers.

6. Khijadia Bird Sanctuary

Location: Jamanagar District

Area: 6.05 sq.kms.

Being situated close to the sea coast, this sanctuary attracts marine as well as fresh water birds. Although small in area, it is probably richer in bird diversity than Bharatpur. The unique location and typical terrain of the area ensures the availability of both fresh as well as seawater in the same area. This sanctuary attracts migratory birds such as flamingoes, pelicans, cranes, plovers, spoonbills, arctic terns, gulls, skimmers, grebes etc.

7. Thol Bird Sanctuary

Location: Mehsana District

Area: 6.99 sq.kms.

This bird sanctuary attracts a large population of ducks, such as shovellers, brahminy duck, pintail, teals, widgeon, cranes and other migratory birds.

8. Porbandar Bird Sanctuary

Location: Junagarh district

Area: 0.09 sq.km.

Located in the middle of the coastal city of Porbandar which also happens to be the birthplace of Mahatma Gandhi. The most peculiar feature of this city which falls in the semiarid drought prone Saurashtra is the terrain of the whole city, which is saucer shaped. Due to this, even during low rainfall years, sufficient water is available for birds. This small lake attracts large concentrations of flamingoes, cranes, egrets, stilt, etc.

9. Gaga Great Indian Bustard Sanctuary

Location: Jamnagar District

Area: 3.33 sq.kms.

A typical grassland habitat, which has great Indian bustard besides other birds like pipits, larks, coursers etc.

10. Kutch Bustard Sanctuary

Location: Kutch district Area: 2.03 sq.kms.

In this Desert thorn scrub and grassland habitat, good concentration of great Indian bustard is reported.

11. Wild Ass Sanctuary

Location: Surendranagar and Kutch District

Area: 4,95370 sq.kms.

One of the largest sanctuaries in the country, represents true desert habitat. It is inhabited by Asiatic wild asses (Equas hemionus khur), one of the seven sub-species found in the world. More than 2,000 wild asses roam over this vast arid habitat alongwith nilgai, blackbuck etc. This sanctuary also has some rare birds like great Indian bustard, houbara bustard, lesser floricans, cranes, pelicans, spoonbills etc. It is an unique ecosystem, spread over a large tract of saline land.

12. Ratanmahal Sloth Bear Sanctuary

Location: Panchmahal district

Area: 55 65 sq.kms.

It is famous for sloth bears, having typical hilly, deciduous vegetation rich in floral biodiversity. The sanctuary falls in the tribal area of the state and it has richest biological diversity in the whole district.

13. Jessore Sloth Bear Sanctuary

Location: Banaskantha District

Area: 180.66 sq.kms.

This is also a sloth bear sanctuary with dry deciduous scrub type of forests. Besides sloth bear, nilgai, wildboar and panther are other major species found here.

14. Jamughoda Wildlife Sanctuary

Location: Panchmahal District

Area: 130.38 sq.kms.

This area is known for its floral diversity and has panthers, wildboars etc. It is believed that till few decades back, the whole district had the same type of floral and faunal diversity.

15. Purna Wildlife Sanctuary

Location: Dangs District

Area: 160.84 sq.kms.

This sanctuary has a typical tiger habitat with moist deciduous forests and is rich in floral and faunal diversity. This sanctuary is characterised by typical western ghat vegetation and it also represents the best forests of western Ghats which falls in the Gujarat state. It is one of the most important gene pool of floral resources and a unique ecosystem of the state.

16. Narayan Sarovar Chinkara Sanctuary

Location: Kutch District

Area: 765.79 sq.kms.

This sanctuary has desert thorn scrub and grassland habitat with excellent concentration of Chinkaras and occasional Bustards. Recently, the area of this sanctuary has been reduced.

17. Shoolpaneshwar Wildlife Sanctuary

Location: Bharuch District

Area: 607.70 sq.kms.

Falling in the catchment area of Karjan and Narmada rivers, Shoolpaneshwar has excellent floral and faunal diversity. Animals include sloth bear, sambhar, wildboar, rusty spotted cat, toddy cat, leopard cat, panther, giant flying squirrels, barking deer, chowsingha and a variety of birds like sun-birds, parakeets, birds of prey, grey hornbill, woodpeckers, etc.

18. Borda Sanctuary

Location: Junagarh and Jamnagar Districts

Area: 192.31 sq.kms.

The habitat is hilly and dry deciduous, located in arid regions of Saurashtra. The rainfall is very low, scanty and erratic, often leading to droughts. The biotic pressure on this sanctuary is very high. The forest is degraded but during the past few years, a concerted effort has been made to conserve and improve the ecosystem. This area has been envisaged as an alternative home of Asiatic lions. The moisture conservation efforts and improved habitat protection have given excellent results. This scrub forest is very rich in floral diversity. Animals reported are sambhar, nilgai, wildboar, panthers etc. This area falls in the Greater Gir Ecosystem along with Girnar, coastal forests and grasslands.

19. Rampara Wildlife Sanctuary

Location: Rajkot District

Area: 15.01 sq.kms.

A small sanctuary with grassland habitat, is mainly inhabited by nilgai besides wildboar, wolf, chinkara, hyaena etc. The terrain is Suacer shaped and scarcity of water is quite common.

20. Kutch Desert Wildlife Sanctuary

Location: Kutch District

Area: 7506.22 sq.kms.

Typical desert habitat with desert flora and fauna such as bustards, nilgai, caracal, chinkara etc.

21. Paniya Sanctuary

Location: Amreli District

Area: 39.63 sq.kms.

It is a part of Gir forests with the same kind of habitat and wildlife as in Gir. In this sanctuary blackbuck, nilgai, asiatic lions, panthers etc are reported.

22. Balaram-Ambaji Wildlife Sanctuary

Location: Banaskantha District

Area: 542.08 sq.kms.

Typical dry deciduous habitat with animals such as nilgai, chowsingha, panther, sloth bear etc., are seen in the area. The area is rich in floral diversity.

23. Hingolgadh Nature Education Sanctuary

Location: Rajkot District

Area: 6.54 sq.kms.

This Sanctuary is famous as a Nature Education Centre with excellent open deciduous type of vegetation. It is a major centre for imparting nature education. The area is characterised by the presence of fauna such as chinkara, reptiles and birds, etc., in semi-arid, drought prone region.

Madhya Pradesh

The State has 11 National Parks, and 33 Sanctuaries. This area of 16,976 sq km encompasses about 5,100 sq km of sal forests, 3,390 sq km of teak forests, 7,518 sq km of mixed forests and 1,300 sq km of grasslands. The protected areas amount to 3.8% of the State's geographical area (4,42,841 sq km). This is 12.5% of the state's total forest area (1,35,396 sq. km).

The Wildlife (Protection) Act, 1972 was enforced in the State since 25.1.1973. A total ban on hunting was also imposed. After the launching of Project Tiger in the early seventies, Kanha National Park was one amongst the first nine tiger reserves constituted in the country. This park has a glorious conservation history and is internationally renowned for its floral and faunal richness. It also has the unique distinction of harbouring the only known world population of the endangered Branderi Swamp Deer (Cervus duvauceli branderi). This species was virtually saved from the brink of extinction and its number has now increased to almost 400 from a precarious 66, two decades ago. Subsequently, four more tiger reserves have also been added, namely, Indravati (Bastar Distt.), Pench (Seoni Distt.), Bandhavgarh (Shahdol Distt.) and Panna (Panna Distt.).

Many endangered species are conserved in the protected areas of the state, like tiger, branderi barasingha, wild buffalo, great Indian bustard, lesser florican, gharial, mugger etc.

Research

Considerable advancements have been made in the field of wildlife research and several research papers have been published under the auspices of Project Tiger in Kanha on the following lines. Monitoring changes in vegetation and abundance of prey-predator population, use of remote sensing techniques (satellite imageries and aerial photographs) for wildlife habitat evaluation & utilisation, grassland management for wild ungulates, translocation of barasingha and the like. Apart from these, species specific research on barasingha, Hanuman langur, tiger and gaur has been done for doctoral research. Separate management plans have been prepared for most of the protected areas in the State.

Special emphasis has been given to ecodevelopment of the villages near the protected areas so as to achieve the twin objectives of socioeconomic upliftment of the local people to elicit their support for wildlife conservation, particularly in Kanha National Park and Nauradehi sanctuary. The following works have been envisaged under ecodevelopment:

- Improvement of agricultural land
- Irrigated pasture development
- Crop protection from wild animals
- Water sources for people & cattle
- Small irrigation schemes
- Well construction
- Use of non-conventional sources of energy

Some important managerial issues that need to be addressed are:

Population pressure vis-a-vis wildlife survival

- Cattle grazing
- Forest fire
- Illicit collection of wood and non-wood products
- Poaching.

1. Strengthening wildlife protection and management in Reserve Forests/Protected Forests and non-forested areas.
2. Control of damage to man and his property by wild animals like crop raiding, cattle lifting, man eating, etc.
3. Control of wildlife diseases.
4. Weed problems.
5. The need of trained wildlife managers and researchers and institutional improvements.

6. The need for larger zoological parks in near natural setting for conservation and breeding of endangered species.
7. The need of wildlife interpretation and extension services.
8. Wildlife based tourism.
9. The broad objectives for future planning could be as follows:
    - Development and protection of forest corridors connecting the wildlife protected areas.
    - Preparing wildlife data base and inventories.
    - Ensuring support of local public for wildlife.
    - Management plans for all the protected areas.

The above objectives would be achieved to a large extent under the Biodiversity component of the World Bank Forestry Project.

North-east India

India's north-eastern region is a unique biogeographical entity. This is a region of physical, ecological, biological, social, cultural and linguistic diversity. Over the centuries, the culture, tradition, belief and life style of people of the region have had close links with nature; many of the communities in the region continue to live very close to nature and even worship it, as manifested in their socio-religious practices and life style.

Shifting cultivators in N.R. India plan their lives on the basis of the cycle of clearing, burning, tilling, planting, weeding, protecting, harvesting and regenerating vegetation in the fallow fields to recover nutrients over the subsequent years before the cycle begins a new. A wide range of crops are grown traditionally in ihcjlium field, thereby transforming the natural forest into a harvestable agricultural land, which does not necessarily loose diversity on a landscape scale, as such fields may contain 8-15 major crops, plus dozens of other minor crops totaling to atleast 35 crops (Ramakrishnan, 1992).

The North-East India is one of the mega biodiversity centres in the Indian subcontinent. It is also one of the 10 distinct biogeographic zones of the country (SAARC, 1992). Its ecosystems vary from tropical wet evergreen, moist deciduous, sub-alpine, alpine forests, and grasslands to numerous freshwater lakes, rivers, swamps and marshy wetlands. These diverse ecosystems are the home of equally diverse range of flora and fauna.

Table 16.1. Status of forests in north-east India (in sq km)

States	*Geographical Forest Area*	*Total Forests Cover FSI 1993*	*Recorded Forests*	*Govt. Forest*	*Community Used/ Controlled Forests*
Arunachal Pradesh	83743	68661	51540	18591	32949
Assam	78438	24508	30708	21520	9188
Manipur	22327	17621	15154	5634	9520
Meghalaya	22429	15769	9496	903	8593
Mizoram	21081	18697	15935	7077	8858
Nagaland	16579	14348	8625	1163	7462
Tripura	10486	5538	6292	6292	-
	255083	165142	137750	61180	76570

Source: Ministry of Environment & Forests, Regional Office, Shillong

One of the best manifestations of biodiversity conservation of the region is perhaps in the form of sacred groves, as seen among the different communities of the region, particularly among the Khasis of Meghalaya. Recently, Tripathi et al. (1995) has documented the existence of 71 sacred groves in Meghalaya, varying in areas from 0.01 ha to 900 ha, and of course, at different degrees of 'sacredness'. Some of these groves are still well preserved even to this day, acting as islands of relic forests and biodiversity.

Table 16.2. Biodiversity/species richness (recorded) in north-east India.

Group	*N.E. India*	*Group*	*N.E. India*
Plants		Algae	
Angiosperms	7500	Fungi	
Orchids	700	Animals	
Bamboo	58	Insecta	3624
Citrus	64	Mollusca	50
Gymnosperms	28	Pisces	236
Bryophytes	500	Amphibia	64
Pterydophytes	700	Reptilia	137
Lichens	728	Aves	541
Mammalia	160		
		Total	15090

(*Source:* BSI & ZSI, Shillong)

The history of the faunal diversity of the region is intimately connected with geomorphological evolution of the area. It is worthwhile mentioning here that the obliteration of pro-tertiary Tethy's sea began in North-East India producing in its wake a land bridge between Indian peninsula and main land Asia. This region served as a faunal gateway and therefore, a complex assemblage of faunal elements are observed in the region (Biswas & Ghosh, 1976).

Located in the north-east of the Indian sub-continent, criss-crossed by Himalayan ranges flanking the Brahmaputra and Surma valleys, the constituent units of the region, i.e. Arunachal Pradesh, Assam, Manipur, Meghalaya, Mizoram, Nagaland and Tripura offer a complex-variety of habitats and ecosystems. The rich forests have greatly suffered due to biotic interference and land exploitation such as shifting cultivation on the hills and rice cultivation in the valleys. This has drastically affected the distribution and abundance of the faunal wealth of the region which, however, still retains numerous unique and rare/endangered species of great scientific and economic value (Ghosh and Tiwari, 1984).

One way of arresting this trend of depletion was through the creation of managed forests and protected areas. In the context of North East India, such managed forests and protected areas constitute about one fourth of the total geographical area of the region.

Fauna

It may undoubtedly be said that the North East India is one of the single richest biogeographic zones in the Indian sub-continent, presenting such an array of richness and unique characteristics of faunal wealth. The faunal resources of North-East India has been described earlier by Ghosh and Tiwari (1984). Their description mainly includes vertebrate Phyla Mammalia, Aves, Reptilia, Amphibia, Pisces and some groups of invertebrates, notably butterflies and beetles. Since then, many more species were described from NorthEast India, particularly the invertebrate groups belonging to Rotifera, Annelida, Crustacea, Cladocera, Myriapoda, Odonata, Orthoptera, Phasmida, Dermaptera, Dictyoptera, Isoptera, Mallophaga, Hemiptera, Diptera, Trematoda, Nematoda and soil arthropods, notably Acari and Collembola. Many of the species recorded from the region were new to science. Some of the recently described new species from North-East India, whose para-types are in the holdings of Eastern Regional Station, Zological Survey of India, Shillong include at least 94 new species belonging to different groups.

It is not that these animals are confined to the managed forests and protected areas of the region alone. In fact, as can be seen from the legal status of the forests in N.E. India (Table 16.1), most of the forest areas, except in Tripura, are under the ownership/control of the community in the form of village councils, district councils or private ownership. Almost all the areas of Unclassed State Forest are under this category and the local tribals enjoy excessive concession in the exploitation of the products from these forests. Faunal endemism is also very high in North-East India. Some of the fauna restricted to North-East India by Darlong et al., 1989; and Alfred, 1990.

The soil fauna is perhaps one of the least known biotic frontiers from this region. But from preliminary studies on the soil fauna of this region, it is observed that the groups promises to constitute a huge reservoir of biodiversity, particularly the two dominant groups Acari and Collembola, whose species richness is also very high.

The soil fauna of some of the managed and protected forests studied from this region by Alfred et al., 1989; Darlong and Alfred, 1993. They show that the soils of these systems are a reservoir of soil faunal diversity. On the contrary, the forest soil fauna could be much richer and denser if sampling is beyond 40 cm (the present studies were limited to 40 cm), a range of extraction methods employed, rather than confining to only Berlese-Tullgren funnel extractions as was the case for the present studies, and careful sorting and identification of all groups and stages could be done.

The effect of disturbance, especially fires resulting from shifting or swidden cultivation, has a profound effect on all animals and especially soil animals. They have reported a decrease in faunal density to the tune of approximately 50 % or more in most groups of soil animals.

Askot Wildlife Sanctuary Of Kumaun (Himalaya)

With an area of approximately 600 sq. kms., ranging between 600-6700 m, Askot Forest Range forms the sanctuary area. An international boundary with Tibet forms its northern limit. The river Kali in the east separates it from Nepal. The boundaries of East Almora Forest Range and Pithoragarh Forest Range constitute the western and southern limits of the sanctuary respectively. The area also includes three major watersheds (Kali, Dhauli and lower part of Gori) of the Kumaun region.

The Askot Wildlife Sanctuary (AWLS) covers most of the area in Askot Forest Range (28943.2 ha). The spatial distribu-

under administrative control of Forest Department. Pinus roxburghi (14%) and Quercus leucotrichophora (14.2%) are the major forest types at lower elevations, whereas Quercus xeinecarpifolia (16%) and Tsuga dumosa (14.9%) are represented at higher elevations. Samant et al. (1993), have reported the detailed general distribution pattern of flora and fauna along the vertical gradient.

Table 16.3. Spatial distribution of forest types and other land-use patterns in AWLS

Forest type	*Area(ha)*	*Per cent*
Pine (Pinus roxburghi)	4043.9	14.0
Oak (Ouercus leucotrichophora)	4123.2	14.2
Tilonj (Q. floribunda)	1189.5	4.1
Kharsu (Q. semecarpifolia)	4631.5	16.0
Blue pine (Pinus wallichiana)	85.4	0.3
Fir (Abies pindrow)	222.9	0.8
Tansen (Tsuga dumosa)	4319.9	14.9
Sal (Shores robusta)	69.2	0.2
Alpine forest (Betula utilis. Rhododendron campanulatum)	2405.2	8.3
Meadows	2292.8	7.9
Barren lands	4162.6	14.4
Inaccessible rocks etc.	1397.1	4.8

(*Source:* Gupta and Chandra, 1977-79 to 1987-88)

A significant population of human and livestock exists in and around the sanctuary. Two tribal groups, viz., Bhotiyas and Rajees, are prominent. In Dharchula Tehsil, constituting the largest administrative unit of AWLS, the trend in the increase of human population is apparent (1981-37,815; 1991-48,704). Irrespective of the socioeconomic status of different communities in AWLS, they depend on wild plant resources of the area to complement their food requirement and/or to generate money. This is in addition to their fodder and fuelwood needs. It is in this context that the use values were determined to strengthen the initial data base on inventory.

Of the total 689 taxa (Gymnosperms, Angiosperms and Pteridophytes), 172 species (21.5%) were found to be utilized by inhabitants of AWLS in one way or the other (Samant et al. 1993). About 72 ethnobotanically important taxa are used locally for complementing food requirements whereas 69 species are used as medicine.

Table 16.4. Number of plant species in various use categories in AWLS

Use categories	*Species (No.)*
Medicinal	69
Food/edible	72
Fodder	57
Cultivated tools	5
House building	9
Fibre yielding	2
Religious purpose	7

Considering the biogeographic elements, it is noticed that 69.2% of the wild useful plants are Himalayan and the rest belong to the neighbouring phytogeographic provinces. The inhabitants of AWLS utilize at least five native Himalayan species (Angelica glanca, Allinm stracheyi, Didymocarpus pedicillata, Pyrus vestita, and Chimonobambusa jaunsarensis) considered sensitive due to their higher rank of rarity. Besides these, a number of non-Himalayan taxa are preferred and promoted by the local inhabitants for agroforestry systems. For example, Baulunia variegata (Kweral), Ficus hispida (Gartimil), F. semcordata (Khastiya), Ridelia montana (Telpa), Celtis tetrandra (Kharik), Diploknema butyracea (Chiura), Grewia elastica (Pharsain), Litsea nwnopetala (Baiganiya).

Grewia elastica (Pharsain), Litsea nwnopetala (Baiganiya). I'luicbe lanceolata (Kaul) etc. Similarly, Woodfordia fruticoxu (Dhaul) is valued as an efficient fuel and in bee keeping.

Vertical distribution indicates the increase in the potential number of wild useful plants with increasing elevation. The distribution of various types of varieties at different elevation zones and deviations from the expected distribution shows that two attitudinal extremes, i.e., below 1500 in and above 2500 m are represented by higher percentage (30.0% and 40.7% respectively) of endangered taxa. There is an increase in endangered taxa, the possible reasons of this phenomenon could be:

The higher concentration of human population at lower altitudes resulting in the increase of pressure on existing resources. Such rarities are usually referred as anthropogenic (Cody, 1986).

The taxa inhabiting alpine/subalpine ranges are subject to extremes of environmental conditions resulting in the occurrence of small populations in relatively smaller patches of suitable habitats. These are called physiologic.

Two Government controlled Institutions, viz., Parvatiya Bhesaj Sangh (PBS) and Kumaun Mandal Vikas Nigam (KMVN) are associated with collection, storage and marketing of the minor forest produces from AWLS. Total export of these products by PBS range between Rs. 12 to 15 lacs per annum. Similar values are expected from KMVN with an involvement of about 2,500 to 3,000 local people (pers. communication PBS, Darchula). The produce is drawn exclusively from natural habitats. Apparently, natural populations of many of these taxa with higher class of rarity and high trade rates may face danger of further depletion if the extraction of resources continues or is expanded.